**HK**Propel ≫ Accessing your HK*Propel* digital product is easy!

T0261891

If it's your first time using HK*Propel*:

1. Visit HKPropel.HumanKinetics.com.
2. Click the "New user? Register here" link on the opening screen to register for an account and redeem your one-time-use access code.
3. Follow the onscreen prompts to create your HK*Propel* account. Use a **valid email address** as your username to ensure you receive important system updates and to help us find your account if you ever need assistance.
4. Enter the access code exactly as shown below, including hyphens. You will not need to re-enter this access code on subsequent visits, and this access code cannot be redeemed by any other user.
5. After your first visit, simply log in to HKPropel.HumanKinetics.com to access your digital product.

If you already have an HK*Propel* account:

1. Visit HKPropel.HumanKinetics.com and log in with your username (email address) and password.
2. Once you are logged in, click the arrow next to your name in the top right corner and then click **My Account**.
3. Under the "Add Access Code" heading, enter the access code exactly as shown below, including hyphens, and click the **Add** button.
4. Once your code is redeemed, navigate to your Library on the Dashboard to access your digital content.

Product: Laboratory Assessment and Exercise Prescription HKPropel Online Video

Student access code: WSAR-06RU-06AE-PBJZ

NOTE TO STUDENTS: If your instructor uses HK*Propel* to assign work to your class, you will need to enter a class enrollment token in HK*Propel* on the **My Account** page. This token will be provided **by your instructor at no cost to you**, but it is required **in addition** to the unique access code that is printed above.

Helpful tips:

You may reset your password from the log in screen at any time if you forget it.

Your license to this digital product will expire **2 years** after the date you redeem the access code. You can check the expiration dates of all your HK*Propel* products at any time in **My Account**.

If you purchased a used book, you may purchase a new access code by visiting US.HumanKinetics.com and searching for "Laboratory Assessment and Exercise Prescription HKPropel Online Video".

For assistance, contact us via email at HKPropelCustSer@hkusa.com.   04-2022

Laboratory Assessment and Exercise Prescription

Jeffrey M. Janot, PhD, ACSM CEP

University of Wisconsin–Eau Claire

Nicholas M. Beltz, PhD, ACSM RCEP

University of Wisconsin–Eau Claire

HUMAN KINETICS

Senior Acquisitions Editor: Michelle Earle; **Managing Editor:** Hannah Werner; **Copyeditor:** Chernow Editorial Services, Inc.; **Permissions Manager:** Martha Gullo; **Graphic Designer:** Dawn Sills; **Cover Designer:** Keri Evans; **Cover Design Specialist:** Susan Rothermel Allen; **Photograph (cover):** vm / E+/Getty Images; **Photographs (interior):** © Human Kinetics; **Photo Asset Manager:** Laura Fitch; **Photo Production Specialist:** Amy M. Rose; **Photo Production Manager:** Jason Allen; **Senior Art Manager:** Kelly Hendren; **Illustrations:** © Human Kinetics, unless otherwise noted; **Printer:** Walsworth

We thank the University of Wisconsin–Eau Claire in Eau Claire, Wisconsin, for assistance in providing the location for the photo and video shoot for this book.

Printed in the United States of America   10   9   8   7   6   5   4   3   2   1

The paper in this book was manufactured using responsible forestry methods.

Human Kinetics
1607 N. Market Street
Champaign, IL 61820
USA

United States and International
Website: **US.HumanKinetics.com**
Email: info@hkusa.com
Phone: 1-800-747-4457

Canada
Website: **Canada.HumanKinetics.com**
Email: info@hkcanada.com

E8549

Tell us what you think!
Human Kinetics would love to hear what we can do to improve the customer experience. Use this QR code to take our brief survey.

CONTENTS

Considering the level of physical inactivity, the prevalence of chronic conditions such as cardiovascular disease and diabetes, and the continued growth of the older adult population in the United States, the need for qualified and competent exercise professionals to design, implement, and lead individuals in exercise programs has never been more important. The authors of this text have a combined 30 plus years of experience teaching and training students in laboratory assessment and exercise prescription for apparently healthy individuals, those who are at risk for chronic disease, and individuals from clinical populations. Over this time of working with students, we have discovered that attention to the following knowledge and skill domains is important in the development of future practitioners in exercise or related fields. Thus, we have designed this book to touch on all these areas.

First, the individual must be able to collect accurate health and fitness data, whether that is pre-exercise screening data to determine whether it is safe to engage in exercise, or physical assessment data to determine body composition, aerobic fitness, muscular flexibility, and so forth. The more precisely this is done, the better sense we have of where our client is at from a health and fitness standpoint to inform our discussions and decisions with issues such as goal setting and exercise program design. Therefore, a sharp focus on skill development in assessment is crucial for the emerging exercise professional. If one's training and experience were to conclude at this step, we would have produced a well-prepared technician with proficient technical skills in assessment and knowledge of how these tests work. However, to develop students into strong practitioners, we must push their development further.

Second, the individual must be able to first understand and then synthesize and interpret this information for their clients. This is what we refer to as the essential "teachable moment," and we must take advantage of it. It is the time when we need to successfully communicate this information so that it is understandable to our client. If the client is better educated and knowledgeable about their current health and fitness status, it is much more likely that they will be able to set realistic and meaningful goals for themselves, with the overarching goal of creating a greater effect on behavior change. Relying on previous or current knowledge from academic courses or other sources is helpful for understanding this information. Combine this theoretical learning with practical experience communicating with classmates and other individuals, and we will have achieved this next step.

Finally, the individual must be able to take all assessment data gathered, along with consideration of the client's goals, and design an individualized exercise program to meet the overall program goals. It is essential that exercise professionals understand the theoretical basis of exercise prescription and the factors (physiological as well as behavioral) that impact each portion of the program design. In doing so, the individual will be better suited to apply these principles for more effective exercise program design.

The idea for this book came about with the goal of producing a resource for students in exercise science and related fields who are developing basic skills in conducting laboratory assessments, analyzing and interpreting data for clients, and applying this information in the design of an exercise prescription. Therein lies the uniqueness of this text: laboratory assessment and application to exercise program design all in one lab experience. Thus, this book can be used as the primary text for an undergraduate exercise science laboratory assessment and procedures course or as a secondary reference text in an exercise prescription and fitness assessment course that includes some laboratory experiences.

Additionally, individuals who are pursuing certification in both health- and fitness-related areas, especially those offered by the American College of Sports Medicine (ACSM), will find the *Laboratory Assessment and Exercise Prescription* text a useful tool when preparing for a certification exam. For this reason, the information provided within this text is closely aligned with the 11th edition of *ACSM's Guidelines for Exercise Testing and Prescription*. Certification and maintenance of these credentials are often required for positions in the health and fitness industry. Certifications such as the ACSM Certified Personal Trainer and ACSM Certified Exercise Physiologist qualify the individual to implement and modify exercise programs for a variety of clients in a safe and effective manner. Secondarily, this text can serve as a manual for exercise professionals looking to review or improve upon their theoretical and practical knowledge base in laboratory assessment and data interpretation in order to become more effective practitioners when working with clients.

Organization of *Laboratory Assessment and Exercise Prescription*

Laboratory Assessment and Exercise Prescription offers a variety of laboratory chapters that are loosely organized into topical sections. The first section involves laying the groundwork for working with clients, and discusses what students will learn and focus on in each lab. Additionally, lab experiences performing basic fitness assessments and preparticipation screening for physical activity readiness will be provided, which is the basis before starting any exercise program.

The next two sections include the assessment of body composition using both field and laboratory tests, the assessment of resting metabolic rate with a discussion on weight management with a primary emphasis on exercise strategies, and the assessment of aerobic fitness employing both submaximal and maximal exercise testing. A review of the principles of aerobic exercise prescription and a case study to apply these principles is also incorporated in the submaximal testing lab chapter. A lab dedicated to the practice and understanding of the ACSM metabolic equations and calculation of energy expenditure and how these can be applied in an exercise program completes the aerobic fitness assessment section.

The next section encompasses the assessment of muscular fitness through three lab experiences: flexibility and range of motion, muscular endurance, and muscular strength. A review of exercise program design for these three components of fitness and case study examples for application of these principles have been integrated into each of these lab chapters.

The final section covers the assessment of select clinical variables: pulmonary function testing, basic electrocardiography, and functional fitness testing. The electrocardiography lab chapter contains learning the skill of preparing the 12-lead electrocardiogram and the learning and practice of interpreting basic cardiac rhythms. The functional fitness testing lab chapter includes a review of and practice administering tests aimed at determining the functional abilities of the older adult client. A case study is also provided for experience in designing a progressive balance program for a client.

Finally, two appendices are included at the end of the text. The first appendix reviews common classes of medications, their action and therapeutic effect, and considerations for how these medications may affect the exercise response. With increased prevalence and pharmacological management of chronic conditions in clients, it is imperative that the exercise professional be familiar with this information. The final appendix reviews basic emergency procedures for exercise physiology labs. Even though the risk of exercise-induced complications is low in most populations, it is important to be prepared to manage an emergency situation should it occur.

Laboratory Experience Features

Each of the lab chapters contains some or all of the following main sections: Purpose of the Lab, Necessary Equipment or Materials, Calibration of Equipment, Procedures, Common Errors and Assumptions, Normative Data and Interpretation, and Exercise Program Design. Some labs also include one or more case studies. The Purpose of the Lab section provides a good theoretical background for understanding the information presented and data gathered in each lab. However, the background section is not a comprehensive review of the topic area; therefore, the reader may need to do additional reading and research for greater understanding of interpretation and application. This text assumes the reader has familiarity with areas such as anatomy, kinesiology, physiology, and exercise physiology that will assist with a deeper understanding of the content covered in each lab.

The Necessary Equipment or Materials, Calibration of Equipment, and Procedures sections are the "bones" of each lab chapter. A list of equipment or materials (e.g., data collection and exercise programming sheets) needed to complete each lab and a discussion of equipment care and calibration (if necessary) are provided. The Procedures section identifies the specific outcomes and goals for each lab, provides clear and easy-to-follow directions and guidelines to independently learn and complete each lab assessment skill, and discusses the necessary equations to solve for specific variables using the data collected. Videos and other images are presented to support skill development and learning in the labs. A discussion of Common Errors and Assumptions will help sharpen your data interpretation skills through furthering your understanding of the theories behind how these assessments work, who they are best suited for, and how sources of error and assumptions might impact results. The Normative Data and Interpretation section provides the information needed for interpretation of your results for each assessment. Finally, certain lab chapters include an Exercise Program Design section that outlines and reviews exercise prescription principles specific to the topic. Case studies are presented in select lab chapters, allowing the reader an opportunity to apply the learned principles.

ACKNOWLEDGMENTS

Writing a textbook from start to finish requires the effort of many people, both center stage and behind the scenes. It also takes years of preparation and development as an author and scholar to undertake and complete a project such as this. We owe a debt of gratitude to many people in our lives who have helped along the way.

Thank you to the Human Kinetics team involved in the development of this textbook: Michelle Earle, Hannah Werner, Gregg Henness, and Jason Allen. It has been an honor to collaborate with each of you and with others at Human Kinetics. A special thank you to Molly, Tommy, Grace, Rachel, and Dani for their assistance in the completion of this textbook.

There is no better coauthor to work with on this book than Dr. Nick Beltz. You are a true expert in the field and a tremendous colleague. Writing this textbook with you has been one of the most enjoyable experiences in my career.

I am forever grateful to my mentors (Doctors John Porcari, Vivian Heyward, Len Kravitz, and Rob Robergs) for their years of support and mentorship and to my special colleagues (Doctors Saori Braun, Marquell Johnson, Christine Mermier, Ann Gibson, Daryl Parker, Lance Dalleck, and Gary Van Guilder), who continue to inspire me to be the best professional I can be.

Thank you to my students, both past and present, for your boundless energy, passion, and curiosity that have sustained me as a teacher and scholar. This book is for you.

My family remains a limitless source of love, strength, and support for me to take on and complete projects such as this textbook. Jill, Zach, and Ben, I couldn't do this job without you. Mom and Dad, I hope you are proud of the person I have become.

—Jeff Janot

Thank you to my mentor, colleague, friend, and coauthor, Dr. Jeff Janot. I could never repay you for the opportunities you've given me and can't think of anyone else with whom I'd rather work. Thank you as well to my mentors, colleagues, and students at University of Wisconsin–Eau Claire, University of Wisconsin–La Crosse, and the University of New Mexico.

None of this would be possible without the unwavering love and support of my wife, Kjirsten. You and our daughters, Baylor and Bennett, are a constant inspiration. A special thank you to my mother-in-law and father-in-law as well for their help during this process.

Finally, I would like to dedicate my contributions in this book to my parents, Warren and Deborah. The persistence, resilience, and hard work required to complete this project are a direct reflection of your lifelong love and guidance.

—Nick Beltz

INTRODUCTION: FOUNDATIONS OF LABORATORY ASSESSMENT

Purpose of Laboratory Assessment

The undergraduate experience of the allied health student provides a breadth of experiences in the classroom, laboratory, and field. Each of these experiences is important in helping young professionals strengthen their skill set to ensure their success as future practitioners and candidates for professional or graduate school. The laboratory activities throughout this book not only provide information and promote lab skill development but are also designed to prepare students across the broad spectrum of allied health professions to work with and alongside patients, clients, students, and athletes. Importantly, the foundational background information, the techniques used to quantify physical parameters, and the data used to interpret the results throughout the laboratory activities abide by the many underlying core competencies in evidence-based practice domains. Evidence-based practice is the application of the most current research to drive clinical practice, interpretation, and decision making (Sackett et al. 1996). The laboratory activities throughout this book are driven by variations of the evidence-based practice core competency domains (Albarqouni et al. 2018). Specifically, impactful lab assessment experiences share common characteristics with clinical and field experiences across the allied health professions. These characteristics include appropriate test selection based on the needs and abilities of the client, proper client education prior to administering the test, prerequisite technical proficiency to conduct an assessment safely and accurately, and interpretation and evaluation of the results for the purpose of program development.

Selecting the Test

The initial phase of laboratory assessment is to select a test based on not only client needs but also their ability level. This part of the assessment requires communication between the test administrator and the client to disclose what physical parameters are needed and why the testing is necessary. For example, a client may present to the laboratory with a goal of increasing muscular strength. There are many laboratory activities throughout this book that can evaluate muscular strength, but which test should you choose? The conversation that you have with your client during the preparticipation screening (lab 1) will give you a better understanding of which test to choose and why. In our example, your client is a 63-year-old female who is recently retired from 35 yr at a desk job. They report having no previous experience with resistance training but are interested in improving muscular strength. Throughout this text you will learn that a one-repetition maximum (1RM) test is the most valid means to measure muscular strength. Validity refers to the ability of a test to measure what it is intended to measure, or test accuracy. With our client, a 1RM test is the reference method because it directly measures maximal strength, but may not be the most appropriate test due to the relative inexperience of the client and the potential safety risks. We sacrifice some level of test validity to choose a more appropriate means for muscular strength assessment. In this case, we may elect to use a 1RM prediction equation using a submaximal workload.

Another example of a reference method is using hydrostatic weighing to determine fat mass and fat-free mass in the two-compartment model of body composition. Hydrostatic weighing is considered a laboratory method in body composition analysis because administering the test requires expensive equipment and a relatively high level of expertise. Like our muscular strength example, it is not always feasible to perform hydrostatic weighing due to resources or training, so we can use a field method test such as skinfold measurements instead. Skinfold measurements are accurate within ±3.5% body fat, a worthwhile sacrifice in validity for the ease of use and inexpensive equipment (ACSM 2021). Some tests require multiple trials within the same testing session, or for the assessment to be conducted periodically to measure change. The reliability of a test indicates the ability of a test to produce consistent results when measured repeatedly. For example, when tracking the progression of our client's core muscular endurance over the course of a 12 wk program, high test reliability allows us to measure confidently at baseline, 4 wk, 8 wk, and 12 wk to monitor training adaptations and make modifications along the way.

Client Education and Building Motivation

Once an appropriate test has been identified, then you will focus on patient education prior to administering the test. The brief discussion leading up to the technical skills will also help establish value in the assessment. Your foundational knowledge surrounding the topic coupled with the information gathered during the health history questionnaire should drive the discussion. In our example, you want to take the opportunity to explain the importance of lower body strength. The labs throughout this book and your background knowledge gained from accompanying courses can teach you about the natural losses in lower body strength and muscle mass during the aging process, and association between lower body strength and disability risk. Furthermore, providing examples of activities of daily living and instrumental activities of daily living that involve lower body strength, such as climbing stairs, getting out of a chair, engaging in leisure activities, or even maintaining the necessary leg strength to play with grandchildren, will establish the importance and value of the assessment for the client.

The strategy contains elements of motivational interviewing, where conversation with the client about their goals and building rapport through listening and client education may promote behavior change (Hall, Gibbie, and Lubman 2012). A complete understanding of motivation and the motivational interviewing approach is complex and beyond the scope of this book; however, meaningful conversations relating theoretical information to your client's goals may establish a personal connection between your client and the assessment. This has the potential to motivate your client throughout the measurement process as well as promote adherence to their exercise prescription.

Technical Proficiency

Now that you have selected the test and educated the patient on relevant background information, you will detail the procedures and conduct the technical portion of the assessment. Some assessments will only require verbal explanation of the testing procedures, while others will benefit from physical demonstration. Consider demonstrating any safety procedures and proper techniques involved during an assessment, when appropriate. For example, your client would benefit from simultaneous verbal and physical demonstration during a sit-and-reach test or push-up test rather than providing a verbal description alone. Portions of some assessments may also benefit from demonstration, such as the submersion and exhalation technique used in hydrostatic weighing.

Your clients will also be performing physical tasks during many of the labs within this book. Be sure to watch your client closely and provide verbal cues to correct form and encourage effort, when necessary. The underlying assumption for many tests is that the client performs the task at a particular level of intensity and technique. Therefore, anything less than correct effort and technique reduces the validity of the test. Laboratory skills also require various levels of practice to gain adequate proficiency. For example, it is recommended that a practitioner conduct the skinfold assessment on 50 to 100 different clients to reach proficiency (Jackson and Pollock 1985). Assessments such as submaximal aerobic testing require less preparation and training compared to the administration of maximal oxygen consumption using a metabolic cart. The goal for proficiency resides with the instructor of the course or the manager of the exercise physiology laboratory. We recommend that students are taught all technical skills in person by the instructor. Students should practice the skills under the supervision of the instructor until they reach a level of proficiency required to practice on peers. It can be helpful to create a skill proficiency assessment rubric to evaluate each student objectively. An example of a complete proficiency assessment for resting blood pressure can be found at the end of this introduction.

Interpretation and Evaluation of Results

Now that data have been collected, it is time to interpret the data and evaluate your client's performance. The data interpretation and evaluation are not only used to tell your client how well they are doing, but should also be used to foster conversation as you and your client reflect on the test performance and how it relates to their goals. This feedback will be considered when designing the exercise prescription portion of the lab activity, assisting you in exercise selection, goal setting, and client motivation.

We interpret results in a few different ways throughout the lab activities in this book. During interpretation, we compare our client's results with normative values. These normative values are population-specific and are established from previous research. Sometimes the norms are displayed as percentiles, while other interpretations involve category ranks. For example, let's say the total distance covered during your client's 6-minute walk test places them in the 70th percentile for their age and sex. Telling your client that they are in the 70th percentile might not be a meaningful interpretation of their performance. In this case, it would be best to explain to your client that they performed in the top 30% for their age and sex, or that they performed better than 70% of individuals in their age and sex group. The other evaluation used throughout the labs is category ranking. For example, your 53-year-old male client completes a Rockport 1-mile walk test to estimate their cardiorespiratory fitness level, and it results in a score of 32.8 mL·kg^{-1}·min^{-1}. Upon further evaluation it is determined that this places them in the "Fair" category. This ranking doesn't necessarily tell you how they compared to other individuals of their age and sex, but rather only implies the level of their cardiorespiratory fitness. A cardiorespiratory fitness level of 32.8 mL·kg^{-1}·min^{-1} would place this client between the 50th and 55th percentile, which is slightly above average when interpreting percentile rankings (ACSM 2021).

Interpreting body composition results against norms presents another unique scenario. For example, the hydrostatic weighing test for your 37-year-old female client resulted in 26.7% body fat. The category ranking for a 37-year-old female with 26.7% body fat would be "Overfat," and the percentile ranking places them between the 25th and 30th percentile (ACSM 2021). Your client may be discouraged if told that they are "Overfat," but they can be told that the recommended body fat percentage range for optimal health is 17% to 26% for females. Focusing on the optimal health range is the most important discussion point in body composition interpretations, while further discussion regarding personal goals, exercise and nutritional habits, and body composition changes may follow.

Exercise Prescription

Many labs throughout this book have exercise prescription components. An exercise prescription is a chance to apply the results of your lab tests to create an individualized program to help your client meet their goals. The general elements that comprise an exercise prescription are frequency, intensity, time, and type, commonly referred to as FITT.

- Frequency refers to the total number of exercise sessions that are to be completed per week. The frequency of exercise sessions per week depends on the type of exercise, intensity of the sessions, and the overall goals of the client. For example, a beginner client may only need to engage in resistance training 2 d·wk^{-1} while a more advanced client could be 4 to 5 d·wk^{-1}.

- Intensity is the metric used to quantify the difficulty of the exercise being performed. Exercise intensity can be quantified in several ways. During resistance training, we can quantify intensity based on a percentage of the maximum weight that the client can complete in a single repetition, known as the 1RM. Aerobic training intensity can be prescribed as a percentage of a client's maximal heart rate or maximal oxygen consumption. We can also use a rating of perceived exertion scale to prescribe exercise during both resistance and aerobic training based on perceptual feedback from a client. Intensity prescription is determined by desired physiological adaptations as well as the goals and time constraints of the client.

- Time, or duration, is the total amount of time prescribed to a single exercise session. In general, time is prescribed in aerobic training over other modes of exercise. Time and exercise intensity are inversely related, where higher intensity exercise sessions are to be completed in shorter sessions, and vice versa. Resistance training is not prescribed for time, but instead for volume. Volume prescription in resistance training is based on the number of exercises completed during each session. Each resistance training exercise has a specific number of total sets and repetitions to be completed within each session based on the intensity, training goals, and experience of the client.
- Finally, a type or multiple types of exercise will be prescribed for each session. Exercise type is also known as the mode, and will vary based on the client's goals, experience, and underlying conditions. For example, aerobic training can be adapted by using a variety of modes. If your client prefers to use the bicycle or elliptical machine instead of walking or jogging, then it may be appropriate to prescribe lower-impact aerobic exercise modes to help them reach their goals. Exercise type can also vary in resistance training prescription. Free weights, machines, suspension trainers, body weight, resistance bands, and kettlebells are all viable resistance training modes to elicit health and performance benefits.

Skill Proficiency Assessment

Following is an example of a resting blood pressure skill proficiency assessment. Points can be assigned to any of the listed items in the Introduction, Assessment, and Evaluation sections. A skill proficiency assessment can provide you and the instructor with a practical and objective way to evaluate your skill development. We also recommend that you gather feedback from your instructor on the proficiency assessment so you can improve on your laboratory skills.

Introduction

☐ Exchange names with the client
☐ Describe the assessment that you will be performing
☐ Define blood pressure
☐ Explain the importance of measuring and monitoring blood pressure
☐ Ask the client about any contraindications, questions, or concerns

Assessment

☐ Position the client properly
☐ Select the correct cuff size
☐ Place the cuff on the upper arm
☐ Check to confirm that the stethoscope head is in the correct position
☐ Place the stethoscope in the antecubital space
☐ Inflate the cuff to an appropriate pressure
☐ Deflate the cuff at an appropriate rate
☐ Ensure accuracy of systolic blood pressure
☐ Ensure accuracy of diastolic blood pressure

Evaluation

☐ Summarize the results
☐ Categorize the results appropriately
☐ Make recommendations based on client goals

Any skill throughout this book can be evaluated using this general structure. Points can be distributed at the discretion of the instructor and these skill assessments can be used by you and your peers to practice. The goal of a proficiency assessment is to ensure that the student possesses a skill level that allows them to conduct the test with a high degree of confidence, validity, and reliability.

Preparticipation Screening for Exercise

Purpose of the Lab

This lab provides opportunities to practice procedures and decision-making skills involved with properly screening a client who is either looking to begin an exercise program or who is wanting to make significant changes to their current regimen (e.g., progress from moderate to vigorous exercise intensity). The main outcome of this process is to ensure that participation in exercise is going to be both safe and effective for your client. Case study examples will be provided for further practice of these concepts.

The general goal of the screening process is to identify clients who require a referral to their healthcare provider for medical clearance prior to engaging in a new exercise program, especially if previously sedentary. Additionally, clients who wish to progress their current exercise program to higher levels of volume and intensity may need medical clearance. Screening is also used to identify clients with either cardiovascular or metabolic diseases who wish to participate in exercise but could do so more safely in a medically based fitness facility, hospital, or clinic and to identify clients who have medical conditions that make exercise unsafe at the present time. This issue is particularly important because they will need referral for further management and control of these conditions until engagement in exercise is deemed safe (ACSM 2021).

Necessary Equipment or Materials

- One copy of the health history questionnaire
- One copy of the informed consent form for exercise participation

- One copy of the exercise preparticipation screening form
- Extra paper for note taking during case study analysis

Procedures

1. Read the following section on normative data and interpretation to gain an understanding of the decision making involved in the preparticipation screening process.
2. Read through and organize the following documents:
   a. Health history questionnaire form
   b. Informed consent form for exercise participation
   c. Medical clearance form
   d. Exercise preparticipation screening form
3. Work with a partner to complete these forms (with the exception of the medical clearance form) and determine their readiness to participate in an exercise program. It is good practice to explain the informed consent form first and then have your client read it and ask questions for clarification if they have any prior to signature.
4. For more practice, read the case studies provided and answer the associated questions.

See the Case Studies for Preparticipation Screening section for more information.

Normative Data and Interpretation

This section includes the specific tables and data needed to appropriately screen clients prior to participation in exercise. These tables will also be used to answer questions contained within the case studies.

Tables 1.1 and 1.2 provide an overview of the information required to identify the initial exercise intensity at which a client should begin if they either do not participate in regular exercise (table 1.1) or participate in regular exercise (table 1.2). In the preparticipation screening process, regular exercise is defined as being performed for 3 d or more per week at a moderate intensity for greater than 30 min and for the last 3 mo or more (ACSM 2021). We will refer to this as the "rule of 3s." When we turn our attention to the identification of cardiovascular disease (CVD) risk factors, different criteria will be used to define insufficient physical activity.

Once this is determined, we move on to the identification of present signs or symptoms that are associated with cardiovascular, metabolic, and renal disease. According to the American College

TABLE 1.1   Recommendation for Initial Exercise Intensity: No Regular Exercise Participation[a]

Variable	Option A	Option B	Option C
Signs or symptoms	No	No	Yes, one or more
Known disease	No	Yes	Disease status: NA[b]
Medical clearance	No	Recommended before engaging in exercise	Recommended before engaging in exercise
Initial exercise intensity recommendation	Light to moderate	Light to moderate	Light to moderate
Progression recommendation	Gradual progression to vigorous if warranted by individual goals of client	Due to known disease, client may progress gradually based on exercise tolerance if warranted	Due to signs or symptoms indicative of disease, client may progress gradually based on exercise tolerance if warranted

[a]Client does not meet the rule of 3s.

[b]Disease status does not matter in this situation because signs and symptoms characteristic of disease are present.

Adapted from American College of Sports Medicine, *ACSM's Guidelines for Exercise Testing and Prescription,* 11th ed. (Philadelphia: Wolters Kluwer, 2022), 35.

of Sports Medicine (2021), common signs and symptoms that are characteristic of these three disease classifications are as follows:

- Pain in the chest (angina pectoris) that is accompanied by arm, jaw, and neck pain, and potentially pain between the shoulder blades
- Shortness of breath (dyspnea) at rest or with mild forms of activity or exertion
- Dizziness or loss of consciousness (syncope)
- Swelling of the calves, ankles, or feet (edema)
- Irregular heartbeat (palpitations) or fast heart rate (tachycardia)
- Dyspnea when in a supine or recumbent position or dyspnea following a short period of sleep (2-5 h)
- Discomfort, pain, or cramping in the buttocks, thighs, and calves during activity or exertion that is relieved by rest (intermittent claudication)
- Presence of a heart murmur
- Increasing or unexplained fatigue with normal activities or exercise

Because some of these signs and symptoms can occur without the presence of underlying disease, they should be contextualized within the circumstances surrounding how they are reported. For instance, if a client states that they have fatigue during exercise or other activities, the exercise professional should follow up with a clarifying question such as "What type of exercise or activity do you experience fatigue with?" If they answer with exercise that includes high-intensity interval training or heavy work around their property, that might make sense as an appropriate response to these activities. However, if they describe that over the last 3 wk, they have been getting more and more fatigued during their daily 30 min walk, that might be cause for concern.

Presence or absence of signs and symptoms of disease—along with identification of known cardiovascular (cardiac, peripheral vascular, or cerebrovascular disease), metabolic (type 1 or type 2 diabetes mellitus), or renal disease—determines what column is chosen (Option A, B, or C) to determine recommendations for medical clearance, initial exercise intensity, and progression of intensity. Pulmonary diseases such as asthma and chronic obstructive pulmonary disease (COPD) do not automatically require medical clearance prior to engagement in exercise, especially if they are well controlled and do not significantly limit exercise (ACSM 2021). As such, they are not included in the list of known diseases. In this case, it is more important to properly screen clients with known pulmonary disease for signs and symptoms related to cardiovascular or metabolic disease prior to exercise programming.

TABLE 1.2 Recommendation for Initial Exercise Intensity: Regular Exercise Participation[a]

Variable	Option A	Option B	Option C
Signs or symptoms	No	No	Yes, one or more
Known disease	No	Yes	Disease status: NA[b]
Medical clearance	No	No: moderate intensity Yes: vigorous intensity[c]	Recommended, thus cease current exercise
Initial exercise intensity recommendation	Continue moderate or vigorous	Continue moderate	Client may re-engage in exercise once cleared
Progression recommendation	Gradual progression to vigorous if warranted by individual goals of client	Due to known disease, client may progress gradually based on exercise tolerance if warranted	Due to signs or symptoms indicative of disease, client may progress gradually based on exercise tolerance if warranted

[a]Client meets the rule of 3s.

[b]Disease status does not matter in this situation because signs and symptoms characteristic of disease are present.

[c]If the client has known disease and their goal is to progress to and engage in vigorous exercise, medical clearance should be acquired.

Adapted from American College of Sports Medicine, *ACSM's Guidelines for Exercise Testing and Prescription*, 11th ed. (Philadelphia: Wolters Kluwer, 2022), 36.

Finally, moderate exercise intensity is defined as 40% to 59% of heart rate reserve (HRR) or oxygen consumption reserve ($\dot{V}O_2R$). It can also be described using metabolic equivalents (METs) [3-5.9 METs] and rating of perceived exertion (RPE) [12-13]. Vigorous exercise intensity is defined as ≥60% of HRR or $\dot{V}O_2R$. It is also described as a MET value ≥6 METs and RPE ≥14 (ACSM 2021). Further discussion and application of exercise intensity prescription methods (i.e., $\dot{V}O_2R$ and HRR) will occur in later laboratory chapters.

Table 1.3 shows the modifiable and nonmodifiable factors that are related to the risk of developing CVD, in particular. Gathering and interpreting this information about your client will help in the design of individualized exercise programs, identify areas needing attention through education and lifestyle modification, and serve as a potential reason for referral to a healthcare provider for further follow-up if the current state of the risk factors warrants such a decision.

When all risk factor data are gathered, it is recommended to sum up the number of modifiable and nonmodifiable risk factors to determine the total number present. Regarding elevated high-density lipoprotein (HDL), it is acceptable to consider this as a negative risk factor, which can cancel out one of the other risk factors associated with CVD because of its anti-atherogenic impact (ACSM 2021). For instance, if your client has three risk factors (e.g., current smoker, high blood pressure, positive family history) associated with increased risk of CVD and elevated HDL, the difference would be two risk factors. Additionally, if risk factor information is missing or has not been gathered at the time of screening, then the risk factor should be counted until such a time that it can be measured and confirmed (ACSM 2021). It should also be stated that while BMI and waist circumference measures are useful tools to quantify high body fatness and its associated risk of CVD, they do not consider body composition (fat mass versus lean body mass) of the individual. In some cases, especially with BMI, these data can be misleading. Therefore, qualified exercise and rehabilitation professionals should consider measuring the percent body fat (%BF) of clients using the methods discussed in labs 3 and 4 of this book (CDC 2020). Collecting this information will often provide a clearer picture of body fatness to determine CVD risk.

TABLE 1.3   Modifiable and Nonmodifiable Cardiovascular Disease (CVD) Risk Factors Plus Their Determining Thresholds

Modifiable risk factors	Thresholds for risk factor determination
Insufficient physical activity (PA) volume (Benjamin et al. 2019; WHO 2020b) MET·min^{-1}·wk^{-1} Moderate intensity exercise Vigorous intensity exercise	<500-1,000 <150 min^{-1}·wk^{-1} <75 min^{-1}·wk^{-1}
Smoking (Benjamin et al. 2019; WHO 2020a)	Cigarette smoker either currently or quit within the last 6 mo. Exposure to environmental tobacco smoke (i.e., secondhand smoke) should also be considered.
BMI and waist circumference (USDHHS 1998) BMI Waist circumference (men) Waist circumference (women)	≥30 kg·m^{-2} >102 cm or 40 in. >88 cm or 35 in.
High blood pressure (Whelton et al. 2018) Systolic blood pressure Diastolic blood pressure	≥130 mmHg and/or ≥80 mmHg Blood pressure should be averaged and confirmed on two or more occasions; taking medications for blood pressure control should be considered as a risk factor (see appendix A).
Blood lipids (Grundy et al. 2019) Total cholesterol Low-density lipoprotein (LDL) High-density lipoprotein (HDL) Non-HDL cholesterol	≥200 mg·dL^{-1} (use if no LDL or HDL available) ≥130 mg·dL^{-1} <40 mg·dL^{-1} for men; <50 mg·dL^{-1} for women ≥160 mg·dL^{-1} Taking medications for cholesterol control should be considered as a risk factor (see appendix A).
Blood glucose (Brannick and Dagogo-Jack 2018) Fasting blood glucose Impaired glucose tolerance* Hemoglobin A$_1$C (HbA$_1$C)	≥100 mg·dL^{-1} ≥140 mg·dL^{-1} ≥5.7%
Elevated HDL cholesterol (Benjamin et al. 2019)	≥60 mg·dL^{-1} This is counted as a negative risk factor.
Nonmodifiable risk factors	**Thresholds for risk factor determination**
Age (Rodgers et al. 2019) Men Women	≥45 yrs of age ≥55 yrs of age
Positive family history (Lloyd-Jones et al. 2004) 1. Sudden death (cardiac) 2. Coronary revascularization 3. Myocardial infarction	Father or first-degree male relative before 55 years old Mother or first-degree female relative before 65 years old

*2 h blood glucose value following a 75 g oral glucose tolerance test (OGTT)

Adapted from American College of Sports Medicine, *ACSM's Guidelines for Exercise Testing and Prescription*, 11th ed. (Philadelphia: Wolters Kluwer, 2022), 47.

Case Studies for Preparticipation Screening

For the case studies, please answer the questions that follow each example. It will be helpful to use the exercise preparticipation screening form to guide you through each case study. The first four questions are very procedural, providing you with practice on the process of preparticipation screening and identifying the exercise intensity at which your client should begin their exercise program. The final question involves the identification of CVD risk factor information that you should address and discuss with your client in a comprehensive exercise and health screening and consultation. Additionally, this question will require you to think critically and will challenge your ability to synthesize and communicate information. The answers to the second part of this question may not be obvious, because there may be many different answers and ways to address the issue at hand. Be creative in your answers and draw from other areas of your academic preparation and training up to this point!

Case Study 1

Demography

> Age: 52
> Height: 6 ft 0 in. (183 cm)
> Weight: 222.2 lb (101 kg)
> Sex: Male
> Race or Ethnicity: White

Family History

Your client's family history reveals the father (age 75), mother (age 72), paternal grandmother (age 92), and paternal grandfather (deceased at age 62) were diagnosed with diabetes and high cholesterol, and a brother (age 48) was diagnosed with high blood pressure at age 38. The father had a coronary artery bypass graft surgery on two coronary vessels (CABG × 2) at age 54, and the paternal grandfather died of a myocardial infarction (MI) at age 62.

Medical History

Present Conditions

A recent physical assessment revealed the following health information. Your client's resting heart rate (HR) was 72 beats·min^{-1}, and resting blood pressure (BP) was 122/88 mmHg (confirmed from a previous measurement). The client's waist circumference measurement was 110 cm (43.3 in.), and a bioelectrical impedance analysis (BIA) showed a %BF of 26%. The blood chemistry panel showed a total cholesterol of 247 mg·dL^{-1}, an HDL of 42 mg·dL^{-1}, an LDL of 168 mg·dL^{-1}, and a triglyceride (TGL) level of 300 mg·dL^{-1}. Fasting blood glucose (FBG) was measured at 114 mg·dL^{-1}. A submaximal aerobic test utilizing a basic step test protocol showed a predicted maximal oxygen consumption ($\dot{V}O_2$max) value of 32 mL·kg^{-1}·min^{-1}. No other data are available at this time.

Past Conditions

Your client has a 15 yr smoking history but quit 2 yr ago and remains a nonsmoker at this time. Otherwise, your client does not report any past medical issues.

Behavior and Risk Assessment

The client works as a branch manager at a suburban bank and undertakes no physical activity during their leisure time. Their healthcare provider, who is concerned about the overall sedentary lifestyle and blood chemistry panel data, has referred this client to you. The healthcare provider prefers trying lifestyle modification before considering medication. Your client used to be very physically active, having played basketball in college and maintained an active lifestyle in the off-season. However, over the past 5 yr, they have become progressively less active because of work and family commitments and is considered sedentary at this time. Your client also reports drinking two or three cups of coffee every morning, one or two soft drinks each afternoon at work, and 10 to 12 alcoholic beverages (mostly beer) every week. A 5 d dietary recall analysis shows a diet of 3,000 to 4,000 calories per day, with a fast-food choice for lunch or supper and no breakfast on most days of the week. Overall composition of the dietary intake shows high amounts of simple carbohydrates and fat followed by protein and low amounts of fruits and vegetables. Please answer the following questions:

1. Does your client participate in regular exercise?
2. Does your client have signs and symptoms (S & S) of disease or known disease?
3. Is medical clearance needed at this time?
4. What is your current recommendation for exercise intensity?
5. What are the client's CVD risk factors (RFs), and what are some ways to reduce CVD risk?

Case Study 2

Demography
Age: 38
Height: 6 ft 3 in. (190.5 cm)
Weight: 230 lb (104.5 kg)
Sex: Male
Race or Ethnicity: Black or African American

Family History
Your client's family history reveals that the father died of an MI at 68 years old, the mother (age 70) had a cerebral vascular accident (CVA) at age 67, and a sister has high cholesterol (age 41 and diagnosed at 35) and obesity.

Medical History

Present Conditions
A recent physical assessment revealed the following health information. Your client had a recent diagnosis of both type 2 diabetes and asthma, which are well controlled by medication (metformin for diabetes and Advair for asthma). They are also taking medication for cholesterol (atorvastatin). Your client's resting HR was 65 beats·min^{-1}, and resting BP was 136/84 mmHg (confirmed from a previous measurement). The client's waist circumference measurement was 101 cm (39.8 in.), and a skinfold assessment showed a %BF of 23%. The blood chemistry panel showed a total cholesterol of 151 mg·dL^{-1}, an HDL of 32 mg·dL^{-1}, an LDL of 101 mg·dL^{-1}, and a TGL level of 214 mg·dL^{-1}. FBG was measured at 122 mg·dL^{-1}, and they report a glycosylated hemoglobin (HbA$_1$C) value of 6.8% from a previous laboratory test. A recent maximal exercise test revealed no cardiac issues during exercise and showed a $\dot{V}O_2$max value of 35 mL·kg^{-1}·min^{-1}. No other data are available at this time.

Past Conditions
Your client was diagnosed with spastic cerebral palsy (hemiplegia) as a child and has fair to good mobility in the upper and lower body. They are a former smoker and quit about 8 mo ago after a 10 yr history of smoking. Otherwise, your client does not report any past medical issues.

Behavior and Risk Assessment
The client is a corporate executive whose healthcare provider has advised them to start getting some more physical activity. Your client was active at a young age and through high school and college, but a very demanding schedule of work, travel, and family obligations (three children under age 10) limit their leisure time. Because your client does not have the time to lift weights and do endurance training like they used to, they have not participated in any regular exercise over the last 6 yr. Your client also reports regular daily caffeine consumption through coffee (one to two cups) and diet soft drinks (one to three 12 oz drinks). Your client reports low alcoholic beverage consumption, mostly

> continued

Case Studies for Preparticipation Screening   *> continued*

during bimonthly social gatherings or functions for work. A 3 d dietary recall analysis shows 2,500 to 3,000 calories per day consumed during three regular meals per day. Most of the meals are prepared at home during the week and contain a balance of carbohydrates (60%), fats (25%), and protein (15%). Weekend "cheat" foods and meals are the main challenge in the overall diet, and calorie consumption tends to go up during this time in the week. Please answer the following questions:

1. Does your client participate in regular exercise?
2. Does your client have S & S of disease or known disease?
3. Is medical clearance needed at this time?
4. What is your current recommendation for exercise intensity?
5. What are the client's CVD RFs, and what are some ways your client can increase physical activity at home and work?

Case Study 3

Demography

Age: 68

Height: 5 ft 4 in. (162.6 cm)

Weight: 175 lb (79.5 kg)

Sex: Female

Race or Ethnicity: American Indian

Family History

Your client's family history reveals no cardiovascular issues, although there is a history of depression in the family. Their father (age 85) and mother (age 90) both passed away from cancer. Your client's two brothers (both age 62) and one sister (age 65) are still living and report no significant health issues.

Medical History

Present Conditions

On the health history questionnaire form, your client states that they have experienced episodes of slight syncope and chest pain during various daily activities over the last month. A recent physical assessment revealed the following health information. Your client's resting HR was 59 beats·min^{-1}, and resting BP was 164/88 mmHg (confirmed from a previous measurement). The client's waist circumference measurement was 86 cm (33.9 in.), and a skinfold assessment showed a %BF of 32%. The blood chemistry panel showed a total cholesterol of 182 mg·dL^{-1}, an HDL of 48 mg·dL^{-1}, an LDL of 121 mg·dL^{-1}, and a TGL level of 200 mg·dL^{-1}. FBG was measured at 98 mg·dL^{-1}. Their last exercise test (6 mo ago), which was a submaximal walking test, showed a $\dot{V}O_2$max value of 28 mL·kg^{-1}·min^{-1}. No other data are available at this time.

Past Conditions

Other than a minor surgical procedure on the left knee 15 yr ago, your client reports no previous medical and health issues.

Behavior and Risk Assessment

Your client is a widowed nonsmoker who lives on a rural property near a small town of 500 people. They have been physically active most of their life in the daily management of the property, which has now been taken over by their grown children, and they remain an independent driver. Your client has noticed that in the past few years they have become weaker, especially in the upper body and back muscles, and that many daily tasks, such as lifting, carrying, and walking, are becoming more demanding. In addition to the work on the property, they engage in 40 to 45 min of moderate physical activity 4 to 5 d per week, a routine that has been maintained for the last 5 yr. Also, during a recent visit to their healthcare provider, they were told that blood pressure, "bad" cholesterol, and BMI had "increased," but they didn't know what that meant. Neither exercise nor diet were discussed, and blood sugars were said to be "OK." Your client's diet is good; meals are prepared at home with appropriate amounts of whole grains, fruits, and vegetables, and they rarely eat outside of the home. However, your client likes to snack on candy and simple carbohydrate treats, and usually does so late at night. From an exercise program standpoint, your client wants to regain strength and make daily tasks easier and possibly increase their fitness capacity (aerobic). Over the years, your client has been very involved in community group activity programs that are delivered in town. Although they are not averse to group activity in town, they would like some of the program to be home-based. Please answer the following questions:

1. Does your client participate in regular exercise?
2. Does your client have S & S of disease or known disease?
3. Is medical clearance needed at this time?
4. What is your current recommendation for exercise intensity?
5. Interpret the CVD RF information for your client so they will understand it, comment on ways they can engage in a home-based exercise program to increase strength and aerobic endurance, and briefly discuss the importance of engaging in group exercise activity.

Case Study 4

Demography

Age: 59

Height: 5 ft 6 in. (167.6 cm)

Weight: 148 lb (67.3 kg)

Sex: Female

Race or Ethnicity: Hispanic or Latina

Family History

Your client's sister passed away recently of sudden cardiac death at age 62, and both parents are living and require home nursing care three times a week. The father had a cardiovascular surgical procedure (ablation) to fix a recurrent arrhythmia problem at age 55, and the mother has hypertension, high cholesterol, obesity, and type 2 diabetes (age of diagnosis unknown).

Medical History

Present Conditions

The health history questionnaire form states that your client currently takes BP (hydrochlorothiazide and perindopril), antiplatelet (clopidogrel), and cholesterol (rosuvastatin) medications to manage the risk of a recurring stroke. A recent physical assessment revealed the following health information. Your client's resting HR was 70 beats·min^{-1}, and resting BP was 120/78 mmHg (confirmed from a previous measurement). The client's waist circumference and %BF were not measured. The blood chemistry panel showed a total cholesterol of 180 mg·dL^{-1}, an HDL of 38 mg·dL^{-1}, an LDL of 105 mg·dL^{-1}, and a TGL level of 110 mg·dL^{-1}. FBG was not measured at this time. Your client has not completed an aerobic fitness test in the last 5 yr. No other data are available at this time.

Past Conditions

Your client is a current smoker and has smoked on and off for the past 30 yr. Past health history also reveals a diagnosis of coronary artery disease (CAD) treated by the placement of three intracoronary stents at age 45, a stroke (CVA), and shoulder surgery to repair a partially torn rotator cuff at age 50.

> continued

Case Studies for Preparticipation Screening   > *continued*

Behavior and Risk Assessment

Your client was recommended for therapeutic exercise programming by their healthcare provider to improve mobility (currently fair), balance (currently poor), and overall ability to live independently following a stroke (CVA) 1 yr ago. There is some residual hemiparesis (weakness) and hemiplegia (paralysis) on the left side, but there are no visual disturbances, memory and mood are good, and issues with aphasia have been resolved for the most part. Your client engaged in both physical and occupational therapy 1 mo following the stroke but has not participated in any formal exercise program for the last 4 mo. They did enjoy exercise when they were younger and participated on club and high school swim teams years ago as a teen and young adult. A main goal is to reach the upper cabinets in the kitchen, be able to walk up and down the basement stairs more easily multiple times a day, and feel "young again" by being able to do other regular daily activities. Based on the occupational and physical therapy poststroke assessments, they need to improve overall muscular fitness and cardiorespiratory fitness. Your client has begun a more complete, heart-healthy diet that was prescribed by the clinical dietitians following the stroke. This diet is low in saturated fat, red meat, and sodium and higher in complex carbohydrates, low-fat dairy, fruits, vegetables, and foods that contain more omega-3 and omega-6 fats. Please answer the following questions:

1. Does your client participate in regular exercise?
2. Does your client have S & S of disease or known disease?
3. Is medical clearance needed at this time?
4. What is your current recommendation for exercise intensity?
5. What are the client's CVD RFs, and what are some ways to maintain or improve independent function at home?

CASE STUDY 1 ANSWER KEY

Regular exercise?	No
S & S or known disease?	No
Medical clearance?	No
Recommendation for exercise intensity?	Light to moderate, progress to vigorous over time
CVD RFs?	Age: ≥45 Family history: Father, coronary revascularization procedure Insufficient PA: Not meeting minimum exercise volume levels BMI/WC: Waist circumference >102 cm (40 in.) High BP: Diastolic blood pressure ≥80 mmHg Blood lipids: LDL ≥130 mg·dL^{-1} (total cholesterol doesn't matter in this case) Blood glucose: ≥100 mg·dL^{-1}

Reducing CVD risk:
The client can address most risk factors (high BP, lipids, waist circumference, blood glucose) with changes to the diet, possibly working with a dietitian to start off. The client needs to start a regular exercise program and work on risk factors this way too (especially increasing HDL) and managing weight and body fat through diet and exercise. The client needs to find ways to de-stress to lower blood pressure with better work–life balance and other stress relievers.

CASE STUDY 2 ANSWER KEY

Regular exercise?	No
S & S or known disease?	Yes, metabolic (type 2 diabetes); asthma is not counted as a known disease in this instance since it is controlled
Medical clearance?	Yes (due to known metabolic disease)
Recommendation for exercise intensity?	Following clearance, light to moderate, progress to vigorous over time
CVD RFs?	Insufficient PA: Not meeting minimum exercise volume levels High BP: Both systolic and diastolic are elevated (≥130/80 mmHg) Blood lipids: On medication to control cholesterol and low HDL Blood glucose: HbA_1C ≥5.7%, ≥100 mg·dL^{-1} and diagnosed diabetes

Increasing PA at home and work:
Get an affordable fitness tracker to track progress and for motivation. Physical activity at work: standing versus seated work if possible; arrange a standing desk workstation. Walking meetings in small groups, taking movement breaks throughout the day, cycle or walking commute if possible, park farther away from building to increase steps, plan exercise opportunities while traveling for work (hotel facilities, walking after work day, etc.). Physical activity at home: include family in physical activity and exercise opportunities, plan physical activity opportunities at children's activities if possible, rearrange work schedule if possible or fit some activity in the morning and some at night around work schedule. Resistance training program could be body weight exercises with bands or a suspension training system to be used at home or while traveling.

CASE STUDY 3 ANSWER KEY

Regular exercise?	Yes
S & S or known disease?	Yes, S & S of disease (syncope and chest pain)
Medical clearance?	Yes (to determine reason for symptoms) and cease exercise
Recommendation for exercise intensity?	May return to previous level of exercise if tolerated following medical clearance; progress exercise as tolerated and meet personal goals
CVD RFs?	Age: ≥55 High BP: Both systolic and diastolic are elevated (≥130/80 mmHg) BMI/WC: BMI ≥30; also consider a %BF of 32% Blood lipids: HDL <50 mg·dL^{-1}

Ways the client can engage in home-based exercise and importance of engaging in group exercise:
As a client who lives in a rural area, it is important to be able to prescribe a home-based exercise program to overcome a potentially common barrier of transportation to and from an exercise facility. Purchasing equipment (if affordable) or borrowing equipment from the community program facility are first steps in engaging in home-based exercise. Hand weights and bands are great modalities to use in a resistance training program, in addition to body weight exercises. Items around the home or property can be used as resistance modalities as well (e.g., cans of food, milk jugs with water in them, broom handles). Exercising 35 to 40 min at 3 to 4 d per week is already a good start, but the addition of a fifth day and pushing to 45 min of moderate intensity exercise could improve endurance more (progress this gradually). It would be important to talk to this client about sedentary time throughout the day. It is possible that the current program is good and all they need to do is limit sedentary time around the house. It is important for this client to continue engaging in group exercise programs due to increased time spent alone as a widow living in an isolated area. Group exercise is great for the older adult because it provides camaraderie, social support and interaction, and psychosocial well-being.

> continued

Case Studies for Preparticipation Screening   > *continued*

CASE STUDY 4 ANSWER KEY

Regular exercise?	No
S & S or known disease?	Yes, known disease (cerebrovascular disease leading to CVA or stroke and coronary artery disease leading to stent placement)
Medical clearance?	Yes, recommended following absence from exercise therapy for the last 4 mo
Recommendation for exercise intensity?	Following clearance, light to moderate intensity, progress as tolerated
CVD RFs?	Age: ≥55 Smoking: Current smoker Family history: Sister died of sudden cardiac death at 62 (≤65 yr in first-degree female relative) Insufficient PA: Not meeting minimum exercise volume levels High BP: On medication (thiazide diuretic) Blood lipids: On medication to control cholesterol and low HDL (statin therapy) Blood glucose: Data unknown, need to confirm to discount it

Ways the client can maintain or improve independent function at home:
Maintaining or improving independent function following a CVA or stroke is very important for stroke survivors. The main focus for improving independence is to improve the function of the left (affected) side enough to match the function of the right (unaffected) side. This can be achieved through a functional resistance training program that focuses on improving activities of daily living and improving range of motion in the upper body and strength in both the upper and lower body. This will help with balance as well. This program can be matched with any home physical and occupational therapy exercises that were given following discharge from both programs. Aerobic exercise should be brought along slowly because of issues with mobility that might make aerobic exercise challenging. Thus, multiple bouts throughout the day and limiting sedentary time would be good first goals for this client until they can perform continuous exercise for 30 min or more. Additionally, recommending that your client join a stroke survivorship support group is important for social interaction, connectedness to others going through similar experiences, and psychosocial well-being.

HEALTH HISTORY QUESTIONNAIRE FORM

Client name _____ Date _____

Address _____

Home phone _____ Work phone _____ Email address _____

Date of birth _____ Age _____

Emergency contact person and contact information _____

Personal physician or other healthcare provider _____

Clinic and contact information _____

Date of last checkup and reason _____

What is your current or usual occupation? _____

Medications

What prescribed medications do you presently take? Why do you take them? Please write your dosage.

Medication	Dosage	Why taken

What nonprescription medicines (over the counter) do you take and why?

Over the counter medicine	Dosage	Why taken

Allergies

Are you allergic to or have you had a bad reaction to any medicines or other substances?   ☐No ☐Yes

If Yes, please list the medicine/substance and the reaction:

_____

Health Habits

Do you currently smoke cigarettes?   ☐ No ☐ Yes   _____ packs/day, _____# years

Have you ever smoked cigarettes?   ☐ No ☐ Yes   _____ packs/day, _____# years

When did you quit? _____

> continued

Health History Questionnaire Form > *continued*

On average, how many cups or cans of caffeine-containing beverages do you consume per day?

_____ Tea   _____ Coffee   _____ Soda

On average, how many alcoholic drinks do you have per day on weekdays? _____

on weekends? _____

Overall, how would you rate your diet? (Circle the appropriate number.)

Unhealthy   Healthy

1   2   3   4   5   6   7   8   9   10

Physical Activity History

Have you ever had an exercise test?   ☐ No   ☐ Yes

If Yes:
Date of test _____

Location _____

If it was abnormal, explain: _____

Are you aware of any physical limitation that would prevent you from exercising regularly?   ☐ No   ☐ Yes

If Yes, please specify: _____

Do you currently exercise on a regular basis?   ☐ No   ☐ Yes

If Yes, how many days per week? _____

How long per session? _____

How many months in a row have you exercised? _____

Type(s) of exercise you enjoy doing: _____

Type(s) of exercise you do not enjoy doing or that cause discomfort: _____

_____

How would you rate your level of fitness?

☐ Poor   ☐ Fair   ☐ Average   ☐ Above Average   ☐ Excellent

How much time do you spend sitting or sedentary during the day (approximately)? _____ minutes

List some personal goals that you want to achieve in the program: _____

_____

Personal History

Check the box if you have ever had the following conditions:

Current signs or symptoms suggestive of disease	Yes	When
Pain in the chest at rest or with exertion		
Abnormal (palpitations) or fast heartbeat		
Heart murmur		
Swelling of the calves, ankles, or feet		
Pain or cramping in the buttocks or lower legs with exertion		
Dizziness or loss of consciousness (fainting spells)		
Shortness of breath at rest or with exertion		
Increasing or unexplained fatigue with normal activities or exercise		
Shortness of breath when lying down or at night		
Past medical conditions	**Yes**	**When**
Heart attack		
Open heart surgery (bypass, valve, etc.)		
Cardiac catheterization with or without angioplasty (stent)		
Congenital heart disease		
Heart failure		
Heart murmurs		
Stroke or cerebrovascular disease		
Pacemaker or implantable cardiac defibrillator		
Heart arrhythmia		
Valve disease (aortic, mitral, etc.)		
High blood pressure		
Rheumatic fever		
Thyroid disease		
Diabetes mellitus (type 1 or type 2)		
Kidney disease		
Liver disease		
COPD		
Asthma		
Chronic bronchitis		
Emphysema		
Osteoarthritis or rheumatoid arthritis		
Osteoporosis		
Low back pain		
Joint pain or swelling		
Other orthopedic problems (bad knees, hips, etc.)		
Emotional disorders		
Anxiety		
Depression		
Other condition(s) not listed:		

> continued

Health History Questionnaire Form   > *continued*

If you answered Yes to any of the preceding questions, please elaborate:

_____

_____

_____

_____

_____

_____

_____

I hereby certify all statements provided by me in this questionnaire are complete and true to the best of my knowledge. Further, I give my permission to the staff member to contact my personal healthcare provider or the program's medical director should there be questions or concerns about information in this health history questionnaire form.

Signature _____ Date _____

INFORMED CONSENT FORM FOR EXERCISE PARTICIPATION

I desire to engage voluntarily in the _____ [your program name]. I have answered the health history questionnaire to the best of my ability.

Exercise assessments will take place at the start and end of the exercise program. Exercise tests such as submaximal aerobic endurance, strength, core endurance, flexibility, body composition, balance, and others will test my overall fitness and functional capabilities. Every effort will be taken to ensure that these activities will be conducted and administered in an appropriate and safe manner and that I am screened properly prior to testing. Adverse events during testing could occur, but the risk is minimal if all guidelines are followed. These adverse events could range from soreness to minor muscle strains, cramping, or bruising to abnormal changes in blood pressure and heart rate and, in very rare instances, abnormal cardiovascular events such as fainting, heart attack, or stroke.

For the exercise program, the activities that I will be given are designed to place a gradually increasing workload on the cardiovascular and musculoskeletal systems and thereby to improve their function. I understand that the reaction of the body to such activities cannot be predicted with complete accuracy. There is a risk of abnormal cardiovascular changes occurring during or following exercise. These changes may include abnormalities of blood pressure or heart rate, ineffective heart function, dizziness or fainting, and in rare instances stroke, heart attack, or even death.

I declare that I intend to use some or all of the facilities, equipment, activities, and services offered by the _____ [your program name]. I understand that part of the risk involved in any activity or program is relative to my own state of fitness and health and is related to the awareness, care, and skill with which I conduct myself in the program. I assume full responsibility for my choices to use or apply, at my own risk, any portion of the information or instruction that I receive.

Before starting the program, I should be aware of abnormal signs and symptoms (e.g., chest pain, extreme fatigue, shortness of breath, muscle pain or cramping, lightheadedness) that alert me to stop exercising. I understand that it is my responsibility to promptly notify the staff if I experience any of these problems.

I agree to learn, monitor, and record, as instructed by the staff, my heart rate and my rating of perceived exertion before, during, and after each session. If desired, I am free to ask for my blood pressure to be checked during these times as well.

I understand that it is my responsibility to report to the staff any changes in medication.

I agree not to leave the exercise area without a cool-down period during which my heart rate returns close to the pre-exercise rate.

I have read this form and understand it. I give my consent freely to engage in an exercise program designed to improve my fitness, health, and functional capacity. Any questions that have arisen have been answered to my satisfaction.

Participant signature _____ Date _____

Staff member signature _____ Date _____

From J. Janot and N. Beltz, *Laboratory Assessment and Exercise Prescription* (Champaign, IL: Human Kinetics, 2023).

MEDICAL CLEARANCE FORM

Dear _____:

Your patient _____ has expressed a desire to voluntarily participate in the _____ [your program name]. Our fitness and therapeutic exercise program involves fitness and functional assessment testing as well as the prescription of a progressive exercise program to improve the overall health and function of your patient. An example of typical assessments that your patient will perform are measures of body composition, submaximal aerobic endurance, muscular flexibility, muscular strength, core endurance, and balance, among others. The exercise program will include aerobic exercise, resistance, flexibility, balance, and other areas of training specific to the needs and goals of your patient. Your patient will be screened thoroughly to begin at an appropriate and safe exercise intensity and volume. All exercise sessions will be supervised by both exercise physiologists and students who are trained in administering such programs.

We have asked that your patient obtain medical clearance from your office before starting this program by checking one of the following categories. By completing the form, you are not assuming any responsibility for how the exercise testing and exercise program are administered in our facility. If you are aware of any medical or other reasons why participation in this program would be contraindicated at this time, please indicate that on this form.

If you have questions or concerns regarding our program, please feel free to contact me _____ _____ [your name and phone number] or one of our program staff members. Thank you for your assistance.

_____ The patient should be safe for your program as outlined.

_____ The patient should be safe to participate, with the following modifications:

_____

_____ I recommend that the patient NOT participate for the following reasons:

_____

Healthcare provider name _____ Date _____

Healthcare provider signature _____

Hospital or clinic and contact number _____

From J. Janot and N. Beltz, *Laboratory Assessment and Exercise Prescription* (Champaign, IL: Human Kinetics, 2023).

EXERCISE PREPARTICIPATION SCREENING FORM

Client name _____ Date _____

Use this form to organize your client's health information data for deciding the initial exercise intensity to begin with and progression strategies within the program. The information gathered from the health history questionnaire can be used to guide you through the preparticipation screening process.

1. Follows the rule of 3s regarding regular exercise participation: _____ Yes or _____ No.

2. List signs and symptoms suggestive of cardiovascular (CV) and metabolic disease (if any):

   _____

   *If any signs or symptoms are checked on the health history questionnaire and listed here, the client should either not be allowed to begin exercise, or exercise should be ceased until medical clearance can be obtained.

3. List known CV, metabolic, or renal disease (if any): _____

   _____

   _____

   *If there is no known disease, then you would not have to seek medical clearance for your client. If there is known disease and your client already follows a regular, moderate intensity exercise program, then medical clearance is necessary only if your client desires to engage in vigorous exercise. If your client does not engage in sufficient physical activity and has known disease, medical clearance should be obtained.

4. Recommendation for initial exercise intensity in the program: _____

   _____

   _____

5. Recommendation for intensity progression in the program: _____

   _____

   _____

6. Modifiable and nonmodifiable CVD risk factors (circle or check all that apply):

Modifiable risk factors	Nonmodifiable risk factors
Insufficient physical activity volume	Age
Smoking	Positive family history
BMI and waist circumference	
High blood pressure	
Blood lipids	
Blood glucose (prediabetes)	
Elevated HDL cholesterol	

Total number of risk factors present: _____

*Remember to subtract one risk factor if your client presents with elevated HDL cholesterol.

From J. Janot and N. Beltz, *Laboratory Assessment and Exercise Prescription* (Champaign, IL: Human Kinetics, 2023).

Basic Fitness Assessment

Purpose of the Lab

Physical fitness is the ability to perform activities of daily living (ADLs) safely without undue fatigue. The components that comprise physical fitness are body composition, cardiorespiratory fitness, muscular strength, muscular endurance, and flexibility. Body composition refers to the components of an individual's body mass, particularly fat and fat-free mass. Cardiorespiratory fitness is the capacity of the lungs, circulatory system, and heart to deliver oxygen to the exercising muscle tissue. Muscular strength is the ability of skeletal muscle to exert maximal force against resistance, while muscular endurance is the ability of skeletal muscle to exert submaximal forces repeatedly. Flexibility refers to the extensible properties of skeletal muscle and other tissues surrounding a joint, allowing a joint to move through a range of motion.

All these fitness parameters can be measured with varying degrees of test validity and reliability. Validity is the ability of an assessment to measure *accurately*, while reliability reflects the ability of an assessment to give *consistent* and *stable* scores across multiple measurements or trials. Assess-ments that require more time, expertise, and expensive equipment have higher degrees of validity and reliability. In other words, the practitioner may choose to sacrifice test validity by selecting tests that involve less skill, time, or equipment to administer.

In many cases, a comprehensive set of data is desired to establish a physiological profile for a client. The basic fitness assessment provides the practitioner with data that effectively indicate client strengths and weaknesses across the spectrum of physical fitness and health. Each of the components of physical fitness and additional ways to assess them will be described in greater detail throughout this laboratory manual.

Body Composition

Body composition can be estimated by using basic anthropometric measurements. Body mass index (BMI) is a clinical metric used to identify obesity and risk for obesity related diseases, most notably type 2 diabetes, coronary heart disease, and hypertension (Must et al. 1999). BMI is calculated by dividing an individual's mass (kg) by height (m^2), but the utility of BMI alone to indicate body fatness has been widely criticized due to its inability to differentiate fat mass and lean tissue as well as fat mass distribution (Ortega et al. 2016). Individuals are classified as either underweight (BMI <18.5 kg/m²), normal weight (18.5-24.9 kg/m²), overweight (25.0-29.9 kg/m²), class I obese (30.0-34.9 kg/m²), class II obese (35.0-39.9 kg/m²), or class III obese (≥40.0 kg/m²) based on their calculated BMI (World Health Organization 2000). Deurenberg, Weststrate, and Seidell (1991) established an equation to estimate body fat percentage (BF%) with a standard error of estimate comparable to methods using more advanced body composition techniques in adults (±4.1%) and children (±4.4%).

Circumference measurements are a useful addition to BMI to further indicate fat mass distribution, body shape, and disease risk. The two most common measurements, waist circumference (WC) and hip circumference (HC), can be easily evaluated with the use of an anthropometric measuring tape. WC can be independently used to evaluate fat mass specific to the abdominal region (visceral fat), while HC is used to measure lower body or gluteofemoral fat distribution. Interestingly, larger WC measurements are strongly related to cardiometabolic disease (Klein et al. 2007), while a larger HC protects against many of the same health complications (Cameron et al. 2020). Due to the opposing health indications when WC and HC are taken individually, dividing WC (cm) by HC (cm) allows for a more robust representation of fat mass distribution by way of the waist-to-hip ratio (WHR). A higher WHR is associated with greater disease risk, and threshold values for risk increase with age and vary between male and female populations.

Cardiorespiratory Fitness and Blood Pressure

Cardiorespiratory fitness (CRF) can be quickly and effectively estimated by using a submaximal aerobic stepping test. The reference criterion for evaluating CRF is maximal oxygen consumption ($\dot{V}O_2$max) and once obtained it can be used to classify fitness level and prescribe individual exercise programs. Detailed in later chapters, it is not always appropriate to perform maximal exercise testing. CRF can be evaluated in a few minutes using minimal equipment. The Queens

College Step test has been shown to produce high valid correlations (r = −0.75) between recovery heart rates and $\dot{V}O_2$max, with a standard error of estimate ±8% (McArdle et al. 1972).

Resting blood pressure is the most common clinical test and can be collected as a helpful metric to indicate cardiovascular health. Blood pressure is the force that blood exerts against the arterial walls during the contraction (systole) and relaxation (diastole) phases of the cardiac cycle. Systolic blood pressure (SBP) is greater because the left ventricle ejects blood into the systemic circulation, exerting a large volume and force on the walls of the arteries. Diastolic blood pressure (DBP) is lower because the myocardium relaxes while blood continues to flow through the recoiling arteries. Chronically elevated resting blood pressure, or hypertension, causes undue stress on the vascular endothelium and strain on the myocardium during ventricular ejection.

There is also an association of elevated resting blood pressure and its derived components (mean arterial pressure and pulse pressure) with increased CVD risk (Whelton et al. 2018). Therefore, it is important to classify resting blood pressure, track changes over time, and implement lifestyle modifications that reduce risk for hypertension and future BP-related CVD outcomes. Following the American Heart Association guidelines, blood pressure is categorized as "normal" with SBP <120 mmHg and DBP <80 mmHg and "elevated" with SBP between 120 and 129 mmHg and DBP <80 mmHg. Hypertension is divided into two subcategories, where "stage 1 hypertension" is SBP between 130 and 139 mmHg or DBP between 80 and 89 mmHg and "stage 2 hypertension" is SBP ≥140 mmHg or DBP ≥90 mmHg (Whelton et al. 2018). It is important to note that the diagnosis of hypertension is only appropriate when measured by a medical practitioner based on the averages of multiple measurements obtained over two or more separate occasions.

Mean arterial pressure (MAP) is the average arterial pressure during a single cardiac cycle and is an indirect measurement of systemic vascular resistance and cardiac output. The equation for MAP [MAP = DBP + 1/3(SBP − DBP)] accounts for the discrepancy in time spent during systole compared to diastole during the average cardiac cycle (DeMers and Wachs 2020). No formal categories to interpret MAP currently exist; however, there is value in monitoring acute and chronic changes in MAP. Acutely, MAP ≥60 mmHg is required to perfuse tissues with oxygen. MAP tracked over time can indicate changes in CVD and mortality risk (Fei 2020). For example, for each 10 mmHg increase in MAP there is a 14% increase in all-cause mortality risk among older populations with SBP hypertension (Domanski et al. 1999).

Pulse pressure (PP) is the difference between SBP and DBP and is a surrogate measure of vascular compliance [PP = SBP – DBP]. Vascular compliance is the ability of a vessel to change volume (diameter) for a given change in pressure and gives an indication of stiffness within the arterial walls. Tracking changes in PP over time can also be useful when monitoring CVD risk because it has been shown that a 10 mmHg increase in PP can increase CVD risk by as much as 20% in Caucasian and Asian populations (Blacher et al. 2000).

Muscular Strength and Endurance

Adequate levels of muscular strength and endurance are needed to engage in many functional ADLs such as carrying groceries and laundry, stair climbing, and standing from a chair. Therefore, it is important to preserve muscular strength and endurance throughout the lifespan to support an independent lifestyle. Measuring strength during joint movement (dynamic) may be desirable and most applicable but it is not always appropriate due to safety considerations, time, and cost of equipment.

Maximal muscular force exertion without joint movement is termed isometric strength and can be quickly evaluated by measuring handgrip strength using a handgrip dynamometer. Isometric handgrip strength has been examined extensively as a metric to track training and rehabilitation efficacy, aging, growth, disease progression, disability, and even mortality risk (Hogrel 2015; Lee and Gong 2020). Furthermore, there is a strong correlation (r = 0.77-0.81) between isometric grip strength and isometric knee extension across individuals 18 to 85 years old (Bohannon et al. 2012).

The reference criterion measurement for handgrip strength is hydraulic handgrip dynamometry; however, the specific model of dynamometer must be considered when interpreting results. For example, Guerra and Amaral (2009) showed a strong correlation (r = 0.83) between hydraulic and spring-loaded handgrip dynamometer models, but the spring-loaded model had significantly lower scores (–3.2 kg) compared to the hydraulic model. Handgrip strength is also used to measure muscular strength as a clinical risk factor for functional disability in older adults, known as dysmobility syndrome (Binkley, Krueger, and Buehring 2013).

Many ADLs require a sustained, albeit relatively small, level of muscular force production. Coupling muscular strength and endurance testing gives the practitioner a better idea of overall muscular fitness. An individual may have a satisfactory level of muscular strength but lack muscular endurance or vice versa. The basic fitness assessment uses a push-up test to assess upper body endurance. In line with other tests in this chapter, the push-up test requires minimal equipment, time, and skill. Muscular fitness discrepancies between males and females underpin the rationale for a modified version in push-up test protocol for females. It has been shown that muscular strength in females is 55% of that in males (Fleck and Kraemer 2004) and muscular endurance specific to the standard push-up is 44% of that in males (Augustsson et al. 2009). Since the standard push-up performance in the Augustsson et al. (2009) study was observed in a fit group of females, the standard push-up could represent an intensity greater than that desired for a muscular endurance assessment in females, and electing to use a modified push-up test may be more appropriate for the general population.

It is worth noting that a partial curl-up test has been traditionally used to measure muscular (core) endurance; however, the Canadian Society for Exercise Physiology (2019) and Liguori and the American College of Sports Medicine (2021) no longer recommend its use in muscular endurance testing.

Flexibility

A certain level of flexibility is needed to carry out ADLs such as putting on shoes or a shirt, reaching for a wallet or purse, and kneeling to garden. Historically, there have been links established between inflexibility and increased injury rate in the hamstrings but the overall body of literature on this topic is largely inconclusive (Liu et al. 2012). Dynamic flexibility refers to the ability of a joint to move through active range of motion by engaging the muscle group antagonist (e.g., contracting the hip flexors to swing the leg through the full range of motion limits the hamstring group) while static flexibility is the measured range of motion as the joint is held in a lengthened position (Iwata et al. 2019).

We typically see declines in flexibility with aging due to a decrease in physical activity and an increase in arthritic joint conditions. Among others, one factor that explains the increased fall risk throughout older adulthood is the negative impact of declining muscle flexibility on walking gait parameters (step length and height) and dynamic balance (Iwamoto et al. 2009; Emilio et al. 2014). Therefore, it is beneficial for an older population to maintain a healthy level of flexibility to improve performance of ADLs and mitigate fall risk.

Since the degree of joint flexibility depends largely on joint structure, flexibility should be assessed on a continuum. Uniplanar joints such as the knee have a greater degree of stability and therefore less mobility, whereas multiplanar joints such as the shoulder have greater mobility and less stability. Even stable uniplanar joints may exhibit unstable characteristics when they have an excess of mobility and can extend beyond normal physiological limits (i.e., hypermobility). Hypermobility is desirable in many sport performance scenarios; however, it may predispose an individual to a variety of injuries such as anterior cruciate ligament tear and joint dislocation (Sundemo et al. 2019; Nathan, Davies, and Swaine 2018).

Necessary Equipment or Materials

Body Composition

- Stadiometer
- Body weight scale
- Anthropometric measuring tape

Cardiorespiratory Fitness and Blood Pressure

- Chair
- Adjustable bench step
- Stopwatch
- Metronome
- Sphygmomanometer
- Stethoscope

Muscular Strength and Endurance

- Chair
- Handgrip dynamometer

Flexibility

- Sit-and-reach box

Calibration of Equipment

Body Weight Scale

1. With the platform empty, ensure that the slider reads "0" on a beam scale. Digital scales may require the scale to be zeroed prior to collecting a weight measurement.
2. Place certified calibration weights on the platform in increments of 50 lb (22.7 kg) and write down any difference between the scale reading and known weight at 50 lb (22.7 kg), 100 lb (45.4 kg), and 150 lb (68 kg).
3. Adjust the difference between the scale reading and known weight value by the scale manufacturer's guidelines.

Scale calibration should be performed yearly, at a minimum, depending on regular use of the scale. If manual adjustments cannot be made on a digital scale, place a label on the scale to indicate the specific weight adjustment that should be made.

Aneroid Sphygmomanometer

1. Remove the pressure tube connection from the cuff to the sphygmomanometer that you are testing.
2. Using a Y-connection adapter, attach the cuff pressure tube to the sphygmomanometer that you are testing and the reference mercury manometer.
3. Place the deflated cuff around a rigid cylinder (recommended ~10 cm in diameter).
4. To check for leaks, inflate the cuff to 280 mmHg and observe if the pressure remains stable.
5. Proceed to inflate the cuff to 300 mmHg. Slowly deflate the cuff and document the difference between the reference mercury column and the aneroid sphygmomanometer every 50 mmHg until reaching 0 mmHg.

6. If the average reading of the aneroid sphygmomanometer is ±3 mmHg different than the mercury column, send it to the manufacturer for recalibration.

Handgrip Dynamometer

1. Secure the handgrip dynamometer so that the handle is parallel with the ground. This can be done by simply holding the dynamometer in the desired configuration.
2. Using string or rope, hang a certified weight evenly at both ends of the grip handle.
3. Complete the measurements in increments of 10 lb (4.5 kg) up to 50 lb (22.7 kg) and record the difference.
4. If the average dynamometer reading and known weights are more than ±10% different, send the dynamometer to the manufacturer for recalibration.

Procedures

Body Composition

Height

1. Without wearing shoes or headwear, ask the client to stand up straight with feet together and flat on the floor, place arms at sides, and look straight forward.
2. The client should take a deep breath and hold. While holding, lower the stadiometer arm and compress any hair to the head. Ensure that the stadiometer arm is level.
3. Take the reading to the nearest 0.1 cm.

Weight

1. Ensure that the client is dressed in T-shirt and shorts, if possible. The client should remove their shoes and empty all the contents of their pockets.
2. With the client standing motionless on the platform, record the weight to the nearest 0.1 kg.
3. Calculate BMI using the following formula:

$$BMI = weight\ (kg)\ /\ height\ (m^2)$$

See table 2.1 for BMI category.

4. Estimate BF% using the sex-specific equation:

$$BF\% = 1.20 \times BMI + age - 10.8 \times sex - 5.4$$

For sex: (men = 0, females = 1)

5. See lab 3 for body fat categorization.

TABLE 2.1   Classification of Weight, Waist Circumference, and Disease Risk Based on Body Mass Index (BMI) and High Disease Risk Values for Waist-to-Hip Ratio in Women and Men

BODY MASS INDEX		WAIST CIRCUMFERENCE AND DISEASE RISK BASED ON BMI*		WAIST-TO-HIP RATIO		
		Women <88 cm Men <102 cm	Women >88 cm Men >102 cm		High risk*	
Category	BMI (kg/m²)			Age	Women	Men
Underweight	<18.5	None	None	20-29	>0.78	>0.89
Normal weight	18.5-24.9	None	None	30-39	>0.79	>0.92
Overweight	25.0-29.9	Increased	High	40-49	>0.80	>0.96
Class I obese	30.0-34.9	High	Very high	50-59	>0.82	>0.97
Class II obese	35.0-39.9	Very high	Very high	60-69	>0.84	>0.99
Class III obese	≥40.0	Extremely high	Extremely high			

*Risk of cardiovascular disease, type 2 diabetes, and hypertension

Note: Waist circumference is measured at the umbilicus when classifying disease risk based on BMI; waist circumference is measured at the natural waist in waist-to-hip ratio.

Adapted from World Health Organization (2000); American College of Sports Medicine, ACSM's Guidelines for Exercise Testing and Prescription, 11th ed. (Philadelphia: Wolters Kluwer, 2022); and Bray and Gray (1988).

For example, if your 47-year-old male client has a BMI of 31.3 kg/m² and a WC of 93 cm, then their risk of cardiovascular disease, type 2 diabetes, and hypertension would be "High." A male client with the same BMI but a WC of 103 cm would have a "Very High" risk of cardiovascular disease, type 2 diabetes, and hypertension. If the 47-year-old male client had a WHR of 1.01, they would be at "High Risk" of cardiovascular disease, type 2 diabetes, and hypertension. A WHR less than 0.96 would not be categorized as "High Risk."

Circumferences

1. Ideally, waist and abdominal circumference measurements are taken when male clients are without a shirt and female clients are wearing only a sports bra. If it is determined that the client is uncomfortable with this guideline, simply lifting and either tying or clipping the shirt up is a viable option.

2. Anatomical locations
   - Waist: narrowest portion of the torso from a frontal plane view; typically, this is between the bottom of the rib cage and the iliac crests
   - Abdominal: level with the umbilicus
   - Hip: widest portion of the buttocks from a side or sagittal plane view

3. Place the zero point of the measuring tape on the side of the client's body, directly on a horizontal plane with the desired anatomical location (figure 2.1). Instruct the client to hold or anchor the measuring tape.

4. Wrap the tape around the body by walking the tape measure around the client and completing a full circumference. Positioning the tape by reaching arms around the client's body may be desired; however, it is not always possible to complete the circumference with a reaching technique when working with larger individuals.

5. Obtain and secure the zero mark on the tape with one hand and the measurement value end with the other hand. Ensure that the tape is horizontal at the landmark site.

6. Instruct the client to abduct both arms to 90 degrees or cross arms in front.

7. If using a Gulick anthropometric tape, apply tension to the tape until the spring-loaded

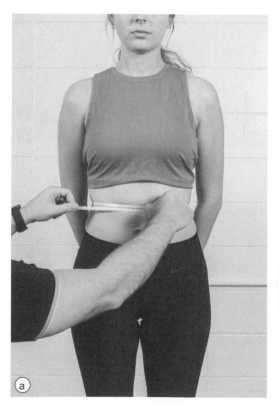
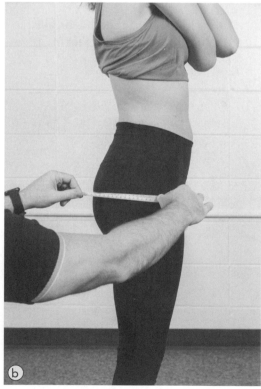

FIGURE 2.1   Waist-to-hip ratio: *(a)* waist circumference; *(b)* hip circumference.

tensiometer marker is reached. If using a measuring tape without a tensiometer, pull the tape taut enough to remove any slack but not so much as to leave an indentation in the skin. Measure to the nearest 0.1 cm.

8. Calculate WHR by dividing the WC by the HC.

9. See table 2.1 for disease risk based on WC, BMI, and WHR.

Cardiorespiratory Fitness and Resting Blood Pressure

Queens College Step Test (McArdle et al. 1972)

1. Set the adjustable bench step to a height of 16.25 in. (41.3 cm). If an adjustable bench step is not available, it is possible to perform this test on any stable surface with a step height equal to the recommendation.

2. Using a metronome, a male client will step at a rate of 24 steps·min^{-1} while a female client will step at 22 steps·min^{-1} for 3 min continuously.

3. After the 3 min, ask the client to remain standing, wait 5 s, and then measure a 15 s heart rate (HR) from 5 to 20 s in the recovery period. It is best to take this measurement at the radial pulse location.

4. Convert HR to beats/min by multiplying the 15 s HR by 4.

5. Estimate $\dot{V}O_2$max using one of the sex-specific equations:

$$\text{Men: } \dot{V}O_2\text{max (mL·kg}^{-1}\text{·min}^{-1})$$
$$= 111.33 - (0.42 \times HR)$$

$$\text{Women: } \dot{V}O_2\text{max (mL·kg}^{-1}\text{·min}^{-1})$$
$$= 65.81 - (0.1847 \times HR)$$

6. See lab 7 for $\dot{V}O_2$max categorization.

 ## Resting Blood Pressure

1. Ensure that the client is seated quietly with feet flat on the floor for at least 5 min prior to the measurement. The client's back and neck should be supported when seated. Remove long sleeves covering the site where the cuff will be placed.

2. Place the client's arm in a supported position such as resting on a table or countertop, fully extended at the elbow. You can also stand and manually support the extended arm. The cuff should remain at the level of the right atrium throughout the measurement.

3. Choose the correct cuff size for the client by ensuring that the bladder within the cuff covers 75% to 100% of the arm circumference. A marking on the inside of the cuff will serve as a guide in most cuffs.

4. Palpate for the brachial artery within the antecubital fossa. Use this landmark as the site for center cuff alignment in accordance with the markings on the bottom edge of the cuff.

5. Wrap the cuff tightly around the upper arm so that the bottom edge of the cuff is approximately 1 in. (2.5 cm) proximal to the antecubital fossa.

6. If using a dual-sided stethoscope, rotate the chestpiece of the stethoscope to ensure that the large diaphragm side is audible. Single-sided stethoscopes may also rotate at the stem to an "off" position, so it is best practice to ensure audible pressure by gently tapping the diaphragm while listening.

7. With eartips pointing forward (aligning with the direction of the ear canals) and inserted in the ears, place the stethoscope diaphragm in line with the brachial artery within the antecubital fossa. Do not allow any part of the stethoscope to touch the cuff. Secure the chestpiece to the arm with your hand and thumb bridging the bell side.

8. Close the inflation valve on the bulb by rotating it clockwise. Squeeze the bulb to slowly inflate the cuff. Listen during cuff inflation and cease inflation once you are 20 to 30 mmHg beyond the point that you no longer hear clear and regular tapping.

9. Open the inflation valve on the bulb by rotating it counterclockwise; the needle on the manometer should drop at a rate no faster than 2 to 3 mmHg per second.

10. Listen carefully for the first appearance of regular tapping. The appearance of regular tapping (two or more consecutive sounds) refers to the first Korotkoff sound and rep-

TABLE 2.2   Classification of Resting Blood Pressure

Category	Systolic blood pressure (mmHg)		Diastolic blood pressure (mmHg)
Normal	<120	And	<80
Elevated	120-129	And	<80
Stage 1 hypertension	130-139	Or	80-89
Stage 2 hypertension	≥140	Or	≥90

Note: Values are an average of at least two separate measurements taken over the course of two visits.
Reprinted from Whelton et al. (2018), p. e21.

resents SBP. Record the value in even digits to the closest 2 mmHg.

11. Continue deflating the cuff and listen for the regular tapping to completely disappear. The complete disappearance of regular tapping refers to the fifth Korotkoff sound and represents DBP. Record the value in even digits to the closest 2 mmHg.

12. Wait 1 to 2 min, repeat the measurement, and average the two measurements. See table 2.2.

For example, if your client's resting blood pressure was 118/76 mmHg then their blood pressure would be "Normal," but if their blood pressure was 118/86 mmHg then their category would be "Stage 1 Hypertension." The systolic and diastolic blood pressures must be interpreted separately and must satisfy the "And" / "Or" criteria to be accurately categorized.

Muscular Strength and Endurance

Handgrip Dynamometry

Using a Jamar Dynamometer (Roberts et al. 2011)

1. The client is seated in a chair. The shoulder is fully adducted and in a neutral position. The elbow is flexed to 90 degrees and the forearm is in a neutral position, resting on the arm of the chair (figure 2.2).

2. Wrist position is between 0 to 30 degrees dorsiflexed, positioned slightly over the end of the arm of the chair with the thumb facing up. Feet are flat on the floor.

3. Position two is the recommended handle position; however, the handle will need to be adjusted for individuals with smaller or larger hand sizes to achieve comfort.

4. Instruct the client to squeeze the handle as hard as they can until you tell them to stop.

5. Encourage effort during the trial.

6. Instruct the client to stop once you see that the gauge needle has stopped rising.

7. Alternate hands and repeat trials until three trials are performed with each hand. Record the score to the nearest 1 kg. The combined highest scores for both hands are used in categorization. See table 2.3.

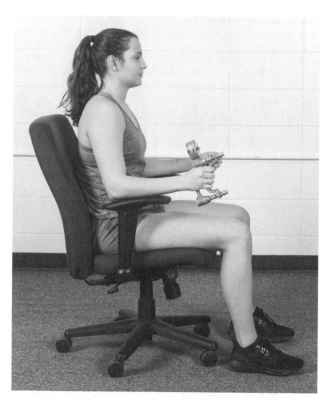

FIGURE 2.2   Grip strength using Jamar dynamometer.

TABLE 2.3 Classification of Grip Strength Score (kg)

	15-19 YR		20-29 YR		30-39 YR		40-49 YR		50-59 YR		60-69 YR	
Category	M	W	M	W	M	W	M	W	M	W	M	W
Excellent	≥108	≥68	≥115	≥70	≥115	≥71	≥108	≥69	≥101	≥61	≥100	≥54
Very good	98-107	60-67	104-114	63-69	104-114	63-70	97-107	61-68	92-100	54-60	91-99	48-53
Good	90-97	53-59	95-103	58-62	95-103	58-62	88-96	54-60	84-91	49-53	84-90	45-47
Fair	79-89	48-52	84-94	52-57	84-94	51-57	80-87	49-53	76-83	45-48	73-83	41-44
Poor	≤78	≤47	≤83	≤51	≤83	≤50	≤79	≤48	≤75	≤44	≤72	≤40

Source: CSEP Physical Activity Training for Health (CSEP-PATH®), 3rd Edition, 2021. Adapted with permission of the Canadian Society for Exercise Physiology.

Using a Spring-Loaded Dynamometer (Canadian Society for Exercise Physiology 2019)

1. While standing, client should grasp the handle of the dynamometer. The grip will need to be adjusted to ensure that the second joints of the fingers are under the handle.

2. The elbow is fully extended. The arm is slightly abducted so that the dynamometer is not contacting the body (figure 2.3).

3. Ask the client to focus on a particular spot on the wall as they maximally squeeze the handle. Instruct the client to exhale as they are exerting effort.

4. Alternate hands until each measurement has been completed twice. Record the highest score for each hand to the nearest 1 kg. The combined highest scores for both hands are used in categorization. See table 2.3.

For example, if your 22-year-old female client's combined handgrip score was 65 kg then they would have "Very Good" muscular strength according to handgrip dynamometry.

Push-Up Test (Canadian Society for Exercise Physiology 2019)

1. Instruct the client to lie on the mat in a prone position with feet together and hands directly under the shoulders.

2. The client will execute a push-up by pressing hands into the floor, moving their body upward until full elbow extension. The upper arms should be internally rotated 45 degrees from the body. Women will complete the test in a modified position, with knees on the ground, acting as the pivot point. Men should perform a standard push-up, with straight legs and the toes acting as the pivot point.

3. From the top position, the client will then lower to the starting position on the floor or mat to complete the repetition. Proper form should be encouraged throughout, where a straight line could be drawn from the back of the heels to the back of the head in men

FIGURE 2.3 Grip strength using spring-loaded dynamometer.

TABLE 2.4 Classification of Push-Up Score (Repetitions)

Category	15-19 YR M	15-19 YR W	20-29 YR M	20-29 YR W	30-39 YR M	30-39 YR W	40-49 YR M	40-49 YR W	50-59 YR M	50-59 YR W	60-69 YR M	60-69 YR W
Excellent	≥39	≥33	≥36	≥30	≥30	≥27	≥25	≥24	≥21	≥21	≥18	≥17
Very good	29-38	25-32	29-35	21-29	22-29	20-26	17-24	15-23	13-20	11-20	11-17	12-16
Good	23-28	18-24	22-28	15-20	17-21	13-19	13-16	11-14	10-12	7-10	8-10	5-11
Fair	18-22	12-17	17-21	10-14	12-16	8-12	10-12	5-10	7-9	2-6	5-7	2-4
Poor	≤17	≤11	≤16	≤9	≤11	≤7	≤9	≤4	≤6	≤1	≤4	≤1

Source: CSEP Physical Activity Training for Health (CSEP-PATH®), 3rd Edition, 2021. Adapted with permission of the Canadian Society for Exercise Physiology.

and a straight line from the back of the knees to the back of the head in women.

4. The test is concluded when the client is no longer able to maintain proper push-up form, visibly strains for two consecutive repetitions, or stops repetitions entirely to rest.

5. The total number of repetitions completed is used for categorization. See table 2.4.

For example, if your 68-year-old female client completed seven total push-ups, then their muscular endurance would be "Good" according to the push-up test.

Flexibility

1. Instruct the client to sit on the floor with knees fully extended and feet 6 in. (15.2 cm) apart and flat against the sit-and-reach box.

2. Ask client to overlap the hands, inhale, and then exhale slowly as they reach forward as far as possible in a slow, controlled movement. Client will hold at maximal distance for 2 s. Knees must remain fully extended and the head should be facing down throughout the duration of the test (figure 2.4).

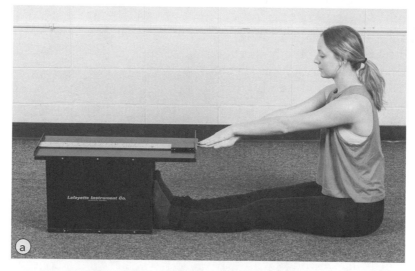

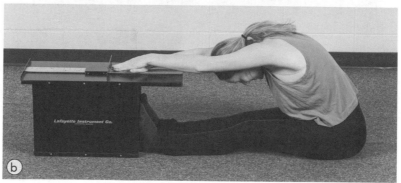

FIGURE 2.4   Sit-and-reach: *(a)* start position; *(b)* finish position.

TABLE 2.5 Classification of Sit-and-Reach Score (cm)

Category	15-19 YR		20-29 YR		30-39 YR		40-49 YR		50-59 YR		60-69 YR	
	M	W	M	W	M	W	M	W	M	W	M	W
Excellent	≥39	≥43	≥40	≥41	≥38	≥41	≥35	≥38	≥35	≥39	≥33	≥35
Very good	34-38	38-42	34-39	37-40	33-37	36-40	29-34	34-37	28-34	33-38	25-32	31-34
Good	29-33	34-37	30-33	33-36	28-32	32-35	24-28	30-33	24-27	30-32	20-24	27-30
Fair	24-28	29-33	25-29	28-32	23-27	27-31	18-23	25-29	16-23	25-29	15-19	23-26
Poor	≤23	≤28	≤24	≤27	≤22	≤26	≤17	≤24	≤15	≤24	≤14	≤22

Source: CSEP Physical Activity Training for Health (CSEP-PATH®), 3rd Edition, 2021. Adapted with permission of the Canadian Society for Exercise Physiology.

3. Record to the nearest 0.5 cm. Repeat the trial and use the highest score for categorization. See table 2.5.

For example, if your 35-year-old male client's highest trial was 18 cm then they would have "Poor" hamstring and lower back flexibility.

Common Errors and Assumptions

Common Errors in Resting Blood Pressure Assessment (Gibson, Wagner, and Heyward 2019)

- Unsupported back, feet, or arm during measurement
- Measurement not taken with arm level with right atrium
- Cuff placed over clothing
- Improper cuff size/alignment
- Improper placement of stethoscope head
- Incorrect side of chestpiece used during measurement
- Thumb placement in the bell
- Deflation rate too fast or too slow
- Background noise (i.e., conversation or music)
- Artifact from client holding on to object such as treadmill handrails or cycle handlebars

Other Considerations

A few conditions exist that may affect blood pressure results, including white coat hypertension and masked hypertension. White coat hypertension occurs when individuals not on hypertensive medication have a normal blood pressure when assessed outside of a clinical setting but develop elevated blood pressure when the measurement is taken by a health professional or in a clinical setting (Franklin et al. 2018). Clients with a history of white coat hypertension should self-monitor their blood pressure at home or in a community-based fitness program. The opposite is true in masked hypertension. Masked hypertension is when an individual has normal blood pressure when measured in a clinical setting but has elevated blood pressure outside of the clinic. Given the relative infrequency that clients have blood pressure measured in a clinic, these individuals do not commonly receive treatment when they would otherwise benefit from it (Pickering, Eguchi, and Kario 2007).

BASIC FITNESS ASSESSMENT

Name _____ Date _____

Age _____   ☐ Female ☐ Male

1. Body Composition

Height _____ cm   Weight _____ kg   BMI _____   Classification _____

Natural waist girth _____ cm   Hip girth _____ cm   Waist/hip ratio _____

Umbilicus girth _____ cm   Evaluation _____

2. Cardiorespiratory Fitness and Blood Pressure

Step test HR _____ bpm   Est. $\dot{V}O_2$max _____ $mL\cdot kg^{-1}\cdot min^{-1}$

Classification _____

Blood pressure: Trial 1 _____ mmHg/_____ mmHg Trial 2 _____ mmHg/_____ mmHg

Avg blood pressure: _____ mmHg/_____ mmHg   Classification _____

3. Muscular Strength

Right: _____ kg _____ kg _____ kg   Left: _____ kg _____ kg _____ kg

Sum of best L + R _____ kg   Classification _____

4. Muscular Endurance

Push-ups _____ reps   Classification _____

5. Flexibility

Sit and reach: Trial 1 _____ cm   Trial 2 _____ cm   Classification _____

Comments

_____

_____

_____

_____

_____

_____

_____

_____

_____

_____

Assessment of Body Composition: Skinfold and Bioelectrical Impedance Analysis

Purpose of the Lab

Body composition refers to the relative percentage of body weight comprised of fat mass (essential and nonessential lipids) and fat-free mass (water, muscle, bone, and other tissues) (Heyward and Wagner 2004). There are no direct methods for determining body composition in humans; therefore, all methods of determining body composition are indirect techniques and will be associated with various assumptions and sources of error that can affect the validity and reliability of these tools.

The ability to determine percent body fat (%BF) and fat-free mass (FFM) has important implications for the maintenance of health and fitness across the lifespan, weight loss programs, growth and maturation, and athletic performance, to name a few. For example, excess body fat has been associated with several chronic diseases including cardiovas- cular diseases such as hypertension and coronary heart disease and metabolic diseases like diabetes, in addition to increasing the risk of benign and malignant cancers (ACSM 2021; CDC 2020). Monitoring growth and maturation of children through body composition assessment is done to determine appropriate progression of body composition or to intervene if there is a health issue (CDC 2021). In addition, competitive and recreational athletes are particularly interested in monitoring their body composition to maximize their overall physical performance (Jackson and Pollock 1985). Measuring %BF and FFM in this population can also be used as a gauge for adaptations to a training program as well as detrimental changes due to overtraining (Heyward and Wagner 2004).

The purpose of this lab is to introduce the assessment of body composition using two common field methods: skinfolds and bioelectrical impedance analysis (BIA). In this lab, you will begin to develop the skills and technical proficiencies necessary to assess body composition using these field methods and learn the main assumptions and sources of error associated with each method. By understanding these underlying issues, the practitioner can collect more accurate body composition data and also better interpret the information for their clients.

Skinfold Method

This method is a noninvasive, indirect technique used to predict one's body composition by measuring skinfolds at specific sites throughout the body. The thickness of the skinfold itself is mostly determined by the layer of subcutaneous fat that resides directly under the skin and has been considered for many years to be a suitable predictor of overall fat mass, lean tissue, and body density (Jackson and Pollock 1978; Keys and Brozek 1953; Lohman 1981; Pascale et al. 1956). Thus, one can get a sense for how much total body fat an individual may have in proportion to the skinfold measurements collected at these various sites (ACSM 2021). Overall, this is an acceptable assumption; however, various factors such as age, sex, race/ethnicity, and overall leanness of the individual can affect how fat is distributed throughout the body and thereby affect the predictive accuracy of the skinfold technique (ACSM 2021; Jackson and Pollock 1985; Lohman 1981; Lohman 1986). To better control for these variations, it is recommended that generalized equations based on sex- or other population-specific equations be used to predict body density and %BF (Jackson and Pollock 1985). Other assumptions and sources of error for the skinfold method will be discussed later in this lab.

The various equations that you will use to determine body composition were chosen based on this recommendation. Interestingly, the equations calculate body density, not %BF, from the skinfold data that you will collect. The body density value will be utilized within the population-specific conversion equations to calculate %BF. When the appropriate techniques/guidelines and equations are used, the predictive accuracy of skinfolds is approximately ±3.5% compared to values obtained through hydrostatic weighing (Heyward and Wagner 2004).

Bioelectrical Impedance Analysis Method

This method is also a noninvasive, indirect technique used to predict one's body composition from measuring the impedance (essentially the resistance) to the flow of electrical current that the analyzer passes through the body (Houtkooper et al. 1996). Resistance, or opposition to current flow, will be lower in tissues (i.e., FFM tissues) with a relatively high content of water and electrolytes, and will be higher in tissues (i.e., fat mass) with a relatively low content of water (Heyward and Wagner 2004). The current of electricity will follow the path of least resistance and will pass much more easily through FFM tissues compared to fat tissue. Therefore, if the measured resistance to current flow is low, it can be assumed that the individual has a lower %BF, and vice versa if the resistance to current flow is high (high %BF).

There is a great deal of variability in the manufacturing and function of BIA analyzers, ranging from research-grade analyzers (e.g., RJL Systems) that use advanced modeling to improve predictive accuracy to those used for general home use to monitor body composition (e.g., Tanita or Omron Body Logic). It is most advantageous to obtain the actual resistance measurement from the BIA analyzer and then use this value in a population-specific equation to calculate FFM and then %BF. However, some models only give measures of either total body water or FFM, or possibly both, and others will give %BF values only. In these cases, it is best to obtain as much information as possible regarding the actual equations and the validity and reliability of these equations used in these models to make the best decision on what to choose for your client (Gibson, Wagner, and Heyward 2019).

When the appropriate techniques/guidelines and equations are used, the prediction error of BIA is approximately within 3.3 to 5 %BF in a diverse adult population compared to values obtained from specific two-component and multicomponent models of body composition, which is very similar to skinfolds (Houtkooper et al. 1996). Two great advantages of BIA over skinfolds are that it is easy to administer from a technical skill viewpoint compared to skinfolds, and is likely more effective at predicting %BF in individuals with obesity due to inherent limitations of skinfold measurements in this population (Gibson, Wagner, and Heyward 2019). However, there are important sources of error and some assumptions that accompany this method that can affect predictive accuracy. These assumptions and sources of error will be discussed later in this lab.

Necessary Equipment or Materials

- Skinfold calipers (Lange calipers recommended)
- Tape measure (Gulick preferred)
- Nonpermanent marker to mark skinfold sites
- BIA machine (or machines if the laboratory has different types)
- One copy of the body composition data sheet

Calibration of Equipment

In general, the accuracy of a skinfold caliper can be measured using a tool of a known width—or better yet, widths—to check across a greater range of measurement values. A calibration block is a tool that is commonly used to check how accurate the caliper jaws are at measuring a given width. Take great care in controlling the caliper during skinfold measurement. Dropping the caliper or snapping the jaws of the caliper shut after skinfold measurement are common factors that can affect caliper accuracy, especially with the high-quality metal calipers. In cases where the skinfold caliper is reading inaccurately, the manufacturer of metal calipers will need to recalibrate them manually, which can be costly. For cheaper plastic calipers, it is best to replace them if they lose their accuracy.

Procedures

The main setup for the lab requires students to work in small groups (2-4 students) to collect data for the skinfolds and BIA. Please record your own individual data, not a group member's data, on the data sheet for later analysis in this lab. Practice taking skinfolds on more than one group member in order to improve your proficiency. Even though a major source of error is intertester reliability (Gibson, Wagner, and Heyward 2019), it is still okay to compare results between each group member and with your lab instructor to have a sense of your ability to measure skinfolds accurately. Other sources of error for both the skinfold and BIA methods will be discussed later.

Skinfold Assessment Guidelines and Procedures

Skinfold Measurement: Chest

Skinfold Measurement: Thigh

Skinfold Measurement: Triceps

Skinfold Measurement: Suprailiac

A detailed description of the standardized process to identify each skinfold site is provided in table 3.1. Read and review these instructions before you start your measurements for this lab. It is extremely important that these guidelines are followed precisely as written and that all sites are identified and marked to reduce measurement error. Please refer to the images in figures 3.1 through 3.9 to visualize the exact locations of these skinfold sites.

To control for measurement error, please follow these standardized guidelines when performing the skinfold technique in this lab (ACSM 2021; Gibson, Wagner, and Heyward 2019; Heyward and Wagner 2004; Jackson and Pollock 1985; Pascale et al. 1956):

1. All measurements should be taken on the right side of the body. Make sure that your client's skin is dry and free of substances (i.e., lotion) that can make skin slippery. This will make the grasping of skin much easier.

2. Follow all standardized site identification procedures as outlined in table 3.1.

TABLE 3.1   Standardized Skinfold Guidelines for the Calculation of Body Density Using the Three-Site and Seven-Site Skinfold Methods

Skinfold site	Standardized guidelines for each measurement
Chest (Jackson and Pollock 1985; Pascale et al. 1956)	Measure diagonally from the anterior axillary (armpit) fold and the nipple. For men, the skinfold is marked and measured halfway along this distance. For women, the skinfold is marked and measured in the first one-third of this distance. The fold is lifted in a diagonal orientation along this line (see figure 3.1).
Abdominal (Jackson and Pollock 1985; Pascale et al. 1956)	Mark and measure exactly 2 cm to the side and at the level of the umbilicus. The fold is lifted in a vertical orientation. The calipers should be on either side of the fold and not contacting the inside edge of the umbilicus (see figure 3.2).
Thigh (Jackson and Pollock 1985)	With the hip and knee flexed at 90 degrees (the client should hold on to a stable support [e.g., table or chair back] and place their foot on a stable surface for good balance and safety), measure along the midline of the thigh and mark the halfway point between the inguinal crease (the junction point between the hip and leg created with the hip flexed at 90 degrees) and the superior border of the patella. Once the site is marked, the client should return their leg to a standing but relaxed position. The fold is lifted in a vertical orientation (see figure 3.3).
Triceps (Jackson and Pollock 1985; Keys 1956; Pascale et al. 1956)	With the elbow flexed at 90 degrees, measure and mark the halfway point between the acromion process of the scapula and the bottom of the olecranon process (elbow) on the lateral side of the arm. Using the tape measure, visualize a line from this point and around to the midline on the posterior side of the arm. Mark this site, ask the client to extend their elbow into a relaxed position, and take the skinfold measurement here. The fold is lifted in a vertical orientation (see figure 3.4).
Suprailiac (Jackson and Pollock 1985)	Identify the anterior axillary fold and, using a tape measure, visualize a line from this position to the iliac crest of the pelvis. Mark the site along this line (anterior axillary line) and slightly superior to the iliac crest. The fold is lifted in a diagonal orientation along the line of the iliac crest and natural tension line of the skin (see figure 3.5).
Subscapular (Jackson and Pollock 1985; Keys 1956; Pascale et al. 1956)	Palpate the inferior angle of the scapula and measure 2 cm below this reference point and mark. The fold is lifted in a diagonal orientation along the natural tension line of the skin (see figure 3.6).
Midaxillary (Jackson and Pollock 1985)	Instruct the client to abduct their arm to 90 degrees to start. Palpate the xiphoid process and identify the midaxillary line (the vertical line that divides the armpit in half). Using the tape measure, visualize a line across from and at the level of the xiphoid process to the midaxillary line and mark this site. The fold is lifted in a vertical orientation along the midaxillary line (see figure 3.7).
ADDITIONAL SKINFOLD SITES COMMONLY MEASURED FOR BODY COMPOSITION ASSESSMENT	
Calf (Gibson, Wagner, and Heyward 2019; Heyward and Wagner 2004)	With the hip and knee flexed at 90 degrees (the client should hold on to a stable support [e.g., table or chair back] and place their foot on a stable surface for good balance and safety), use a tape measure to identify the maximal circumference point of the calf muscle. Mark the skinfold site at this level on the medial side of the lower leg. The fold is lifted in a vertical orientation (see figure 3.8).
Biceps (Gibson, Wagner, and Heyward 2019; Heyward and Wagner 2004)	Using the tape measure, visualize a line from the triceps skinfold site to the anterior portion of the arm directly over the biceps muscle and along the midline of the upper arm. Measure 1 cm above this position and mark this site. The skinfold will be taken at this site and lifted in a vertical orientation (see figure 3.9).

Note: All skinfolds should be taken on the *right side* of the body. The caliper jaws should be placed *directly* on the marked site and approximately 1 cm below the fingers for all measurements.

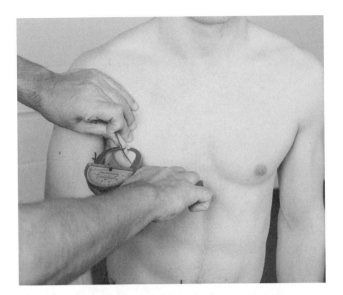

FIGURE 3.1   Chest skinfold.

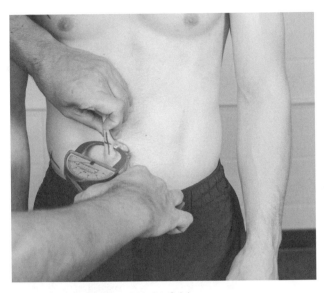

FIGURE 3.2   Abdominal skinfold.

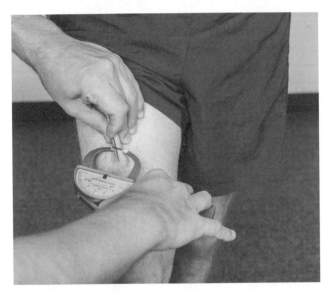

FIGURE 3.3   Thigh skinfold.

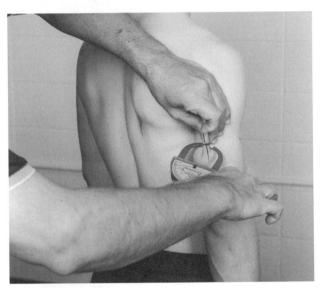

FIGURE 3.4   Triceps skinfold.

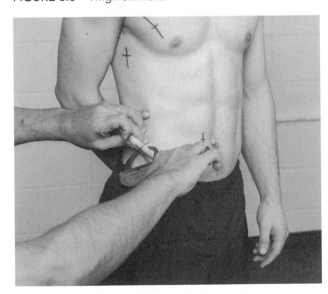

FIGURE 3.5   Suprailiac skinfold.

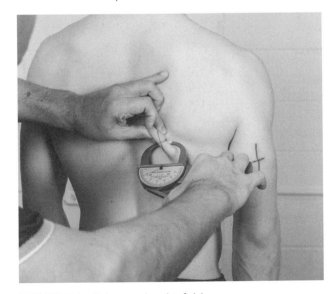

FIGURE 3.6   Subscapular skinfold.

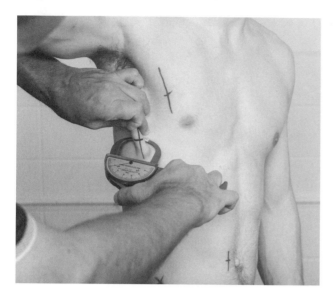

FIGURE 3.7 Midaxillary skinfold.

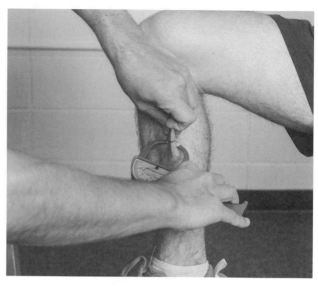

FIGURE 3.8 Calf skinfold.

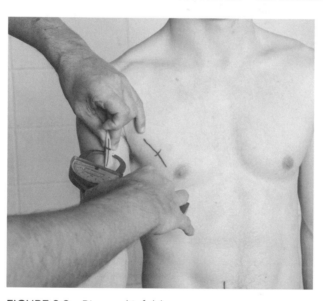

FIGURE 3.9 Biceps skinfold.

3. To lift the skinfold, splay your left hand open, with your thumb pointing down and lay it on your client's skin. Slowly move the fingertips of your thumb, middle finger, and ring finger across the skin and grab the skin as you move them. Hold the skinfold firmly between these fingers approximately 1 cm above the marked site where you will be placing the skinfold caliper jaws. Use the pads of your fingertips to grasp and hold the skinfold, and not the tips, to avoid pinching your client's skin with your nails. It is recommended to keep nails trimmed if planning to measure skinfolds.

4. To ensure that you are grasping skin and fat only, it is good practice to instruct the client to contract the muscle under the skinfold. Keep a grasp on the skinfold while the client is performing this maneuver and allow any muscle tissue to move under and away from your fingers. After this is done, recheck that you have a good grasp on the skinfold and move on to step 5.

5. Position the jaws of the caliper perpendicular to and in the middle (halfway between the base and top) of the skinfold. DO NOT let go of the skinfold while the calipers are in contact with the skin.

6. Slowly release the jaws so that they are fully pressed against the skinfold. Always maintain control over the calipers with a firm grip, keeping your thumb in light contact with the caliper jaw lever. It is acceptable to take a reading of the caliper measurement approximately 2 to 4 s after placement. However, please note that ACSM recommends reading the caliper within 1 to 2 s of placement (ACSM 2021). Thus, if taking an ACSM certification exam, this would be the timing to know.

7. Record the skinfold measurement on your data sheet to the nearest 0.5 mm if using Lange calipers or to the nearest 1 mm if using plastic calipers.

8. After taking the measurement, open the caliper jaws, remove from skin, and slowly close the jaws while releasing the skinfold.

9. Take single measurements at each site and rotate through each site (in order) for a minimum of two times. It is acceptable practice to average two skinfold values if they vary less than ±10% (Harrison et al. 1988). If the values fall outside of this range, then repeat until agreement is reached and average those values. However, please note that ACSM recommends that skinfold values should vary less than 1 to 2 mm as the range for agreement (ACSM 2021). Thus, if taking an ACSM certification exam, this would be the range to know.

Body Density From Skinfold Measurement and Body Density Converting Equations

The following equations are to be used to determine body density from both the three-site and seven-site methods and for converting body density into %BF. The three-site and seven-site equations are sex specific, and body density converting equations are sex and population specific; thus, make sure that you choose the correct equation for your own body composition analysis.

Once the body densities are determined, you will then choose a body density converting equation from the list provided and calculate your %BF for both skinfold methods. When entering the body density into the converting equations, make sure to round the number to five decimal places

(Gibson, Wagner, and Heyward 2019). For example, if you calculate a body density via skinfold analysis that is 1.02345678, that number should be rounded up to 1.02346. Remember, small changes in body density will equal relatively larger changes in %BF; thus, we need to be much more precise with the calculation and use of body density in determining body composition. Once your %BF numbers are calculated, compare the values from each skinfold method to determine how close they are to one another. Past research has identified that these two skinfold methods are highly related to each other in the general population (Jackson and Pollock 1978). Additionally, you can compare your %BF values to the normative chart in table 3.2 to determine where you rank against others of similar age and sex. Finally, please calculate your fat mass and FFM using your %BF value. Simply multiply your body mass in kilograms by the %BF value in fraction or decimal form (i.e., 15% = 0.15). This will give you your total fat mass. Take this value and subtract it from your body mass to get the FFM. Record all values and interpretations on the body composition data sheet for both three-site and seven-site methods.

Three-Site and Seven-Site Body Density Equations

The following equations are to be used for men between the ages of 18 and 61:

Three-site equation (Jackson and Pollock 1978): Chest, Abdomen, and Thigh

Body density $(g \cdot cc^{-1})$ = 1.10938 − 0.0008267 (sum of 3 skinfolds) + 0.0000016 (sum of 3 skinfolds)^2 − 0.0002547 (age)

Seven-site equation (Jackson and Pollock 1978): Chest, Abdomen, Thigh, Subscapular, Midaxillary, Suprailiac, and Triceps

Body density $(g \cdot cc^{-1})$ = 1.112 − 0.00043499 (sum of 7 skinfolds) + 0.00000055 (sum of 7 skinfolds)^2 − 0.00028826 (age)

The following equations are to be used for women between the ages of 18 and 55:

Three-site equation (Jackson, Pollock, and Ward 1980): Triceps, Suprailiac, and Thigh

Body density $(g \cdot cc^{-1})$ = 1.0994921 − 0.0009929 (sum of 3 skinfolds) + 0.0000023 (sum of 3 skinfolds)^2 − 0.0001392 (age)

Seven-site equation (Jackson, Pollock, and Ward 1980): same as men's seven-site

Body density (g·cc⁻¹) = 1.097 − 0.00046971 (sum of 7 skinfolds) + 0.00000056 (sum of 7 skinfolds)² − 0.00012828 (age)

Population-Specific Body Density Converting Equations

Equations to convert body density (Db) to %BF were developed by considering the relative contributions of an individual's fat mass combined with the relative contributions of water, protein, and bone mineral to the FFM. Because body density can be affected by age, sex, race, and health status, to name a few, many different equations have been designed to meet these population-specific factors for a more accurate assessment of %BF. The equations that we will be using in this lab take into consideration many of these individual factors. Therefore, choose the equation that is best related to your own personal background. Finally, population-specific equations do not exist for all racial, ethnic, age, and clinical groups. Thus, in cases where you cannot find a specific equation that meets the criteria for your specific client (or yourself), the best practice at this time is to choose the conversion formulas designed for white men and women (Gibson, Wagner, and Heyward 2019).

The following equations are to be used for white men and women (ages 18-59):

Men	Women
%BF = (495/Db) − 450 (Siri 1956)	%BF = (496/Db) − 451 (Heyward and Wagner 2004)
	%BF = (506/Db) − 461 (Lohman et al. 1984)

The following equations are to be used for Hispanic/Latino men (ages 18-59) and Hispanic/Latina women (ages 20-40):

Men	Women
%BF = (495/Db) − 450 (Siri 1956)	%BF = (487/Db) − 441 (Heyward and Wagner 2004)

The following equations are to be used for Black or African American men (ages 19-45) and women (ages 24-79):

Men	Women
%BF = (486/Db) − 439 (Wagner and Heyward 2001)	%BF = (483.2/Db) − 436.9 (Ortiz et al. 1992)

TABLE 3.2 Percent Body Fat Ratings for Adults by Age and Sex

Age (yr)	Lean to very lean	Good to average	Average to below average	Obese
MEN				
20-29	≤11	11.1-17.1	17.2-23.4	≥23.5
30-39	≤15.4	15.5-20.4	20.5-25	≥25.1
40-49	≤18	18.1-22.4	22.5-26.7	≥26.8
50-59	≤19.8	19.9-23.9	24-28.2	≥28.3
60-69	≤20.6	20.7-24.6	24.7-28.9	≥29
70-79	≤20.6	20.7-24.3	24.4-28.1	≥28.2
WOMEN				
20-29	≤17.2	17.3-22.2	22.3-28.7	≥28.8
30-39	≤17.9	18-22.3	22.4-29.7	≥29.8
40-49	≤20	20.1-26	26.1-32	≥32.1
50-59	≤22.9	23-28.8	28.9-33.9	≥34
60-69	≤23.9	24-29.7	29.8-34.5	≥34.6
70-79	≤23.2	23.3-28.2	28.3-33.7	≥33.8

Adapted from The Cooper Institute (2013).

The following equations are to be used for American Indian men (ages 18-62) and women (ages 18-60):

Men	Women
%BF = (497/Db) − 452	%BF = (481/Db) − 434
(Heyward and Wagner 2004)	(Heyward and Wagner 2004)

BIA Assessment Guidelines and Procedures

As stated previously, since there are a variety of different models of BIA analyzers, this lab will introduce the basic guidelines and procedures for the measurement of body composition using BIA. For this lab, make one (or two if more than one BIA analyzer is available) assessment of %BF using a BIA analyzer and compare the value(s) to the skinfold data and determine the level of agreement. Before any BIA measurement takes place, it is important for the client to follow these guidelines as closely as possible to ensure an accurate assessment result (Gibson, Wagner, and Heyward 2019; Houtkooper et al. 1996):

- The room should be set at a thermoneutral temperature (24-26 °C or 74-79 °F).
- No moderate to vigorous exercise within 12 h of the assessment.
- No consumption of food or drink (except small amounts of water to maintain hydration status) within 2 to 4 h and alcohol within 48 h.
- Remove as much jewelry as possible prior to the assessment. Clients who have indwelling metal devices or objects (i.e., pacemakers, metal rods, etc.) should find an alternative body composition assessment method.
- Empty bladder and bowels within 30 min of the assessment.
- Do not take diuretics prior to the assessment unless directed by a healthcare provider to do so.
- Women who are in the menstrual cycle phase associated with high retention of water should avoid testing at this time.

The traditional BIA method measures total body resistance using electrodes placed on the right side of the body at the wrist and hand and ankle and foot. If you are employing this method, it is important to follow these guidelines (Gibson, Wagner, and Heyward 2019; Houtkooper et al. 1996):

- The client is in a supine position and not in contact with anything metal.
- Clean the dorsal (top) side of the wrist and hand plus the dorsal side of the ankle and foot with an alcohol pad.
- Place an electrode at the wrist so that the top border bisects the head of the ulna and radius and at the ankle so that the top border bisects the lateral and medial malleoli. Do not touch the sticky surface of the electrode or the skin after cleaning.
- Place the other electrodes 5 cm from the wrist and ankle electrodes. You should have a total of four on the skin. Attach the BIA wires to the electrodes according to manufacturer guidelines.
- Instruct your client to remain still during measurement. The client's arms should not be touching the sides of their chest and their legs should not be touching each other.

The bioimpedance spectroscopy (BIS) method measures total body resistance by passing multiple frequencies of electrical current through the body to determine body composition (Gibson et al. 2008). The overall procedures to follow for the InBody BIA device (figure 3.10) are fairly simple. It is important to clean the feet with alcohol pads prior to stepping on the scale and make sure that the hands are clean and dry. Arms should not be in contact with the sides of the chest, legs should not be touching, and the client should remain still during measurement. The client should stand directly over the placements for the feet and maintain a firm grip on the handles. Enter the client data on the screen prior to measurement and record the %BF. The InBody will also provide a measurement of total body water, and some models will also report a value for visceral adiposity.

Two popular BIA analyzers, the Tanita and the Omron Body Logic, are simple and less expensive models that pass a single frequency of electrical current either through the lower body from foot to foot (Tanita) or the upper body from hand to hand (Omron Body Logic). Like with the InBody, the overall procedures to follow for these machines

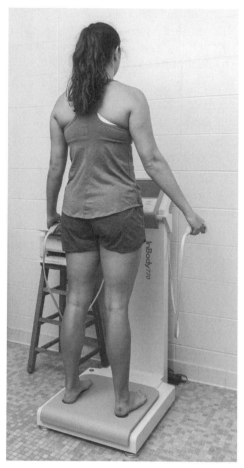

FIGURE 3.10   InBody BIA device.

are very simple. The hands should be clean and dry for the Omron and the feet should be cleaned with an alcohol pad prior to using the Tanita. The arms should not be in contact with the sides of the chest (Omron) and the legs should not be touching (Tanita). Enter the client data into each machine prior to measurement and record the %BF.

Common Errors and Assumptions

Since the overall assessment of body composition involves noninvasive methodologies, there are many assumptions that must be considered and addressed that will affect the validity of the data collected. Likewise, introducing unnecessary error into an assessment method that already contains an inherent level of error in the prediction of body composition will significantly affect the validity and reliability of the tool and data gathered. Thus,

care needs to be taken to meet all assumptions and minimize extraneous sources of error as much as possible. The following sections address both the main assumptions and sources of measurement error for both methods.

Assumptions and Sources of Error in Skinfold Assessment

The best assumptions of the skinfold method are that skinfolds are a good assessment of subcutaneous fat; therefore, when taken at different sites on the body, they can be a reasonable estimate of %BF, and summing skinfolds from standardized sites can predict overall body density (Heyward and Wagner 2004). In contrast, a poor assumption that can affect the accuracy of the skinfold method is that fat is distributed evenly internally and subcutaneously for all individuals by sex. Factors that invalidate this assumption were discussed earlier in the Purpose of the Lab section.

The largest source of error for the skinfold technique is technician skill (Gibson, Wagner, and Heyward 2019; Jackson and Pollock 1985; Pollock and Jackson 1984). Skill development begins with following and practicing the standardized procedures as outlined in table 3.1. Learning from an experienced skinfold technician, practicing your technique on a variety of individuals (i.e., young versus old, lean versus overweight, men versus women, etc.), and gaining experience on the recommended number of clients (50-100 individuals) can increase proficiency in skinfold measurement (Jackson and Pollock 1985; Pollock and Jackson 1984). Thus, in order to increase your skill, you must practice beyond this lab and on other individuals besides your classmates.

Other sources of error include the type of caliper used, equations chosen to calculate body density, and individual client factors. In general, metal calipers such as the Harpenden or Lange skinfold calipers are considered to be the most accurate (Jackson and Pollock 1985). However, some plastic calipers compare favorably to metal calipers and can be less expensive alternatives (Gibson, Wagner, and Heyward 2019).

Compared to BIA measurement, the skinfold method is less affected by individual client factors such as hydration status, although it is good practice for clients to be measured in a euhydrated state to minimize error (Jackson and Pollock 1985).

Finally, compressibility of the skinfold can vary from client to client and skinfold site to skinfold site, which could affect the predictive accuracy of the method in some clients, especially those with higher levels of body fat. If this is the case, it may be best to use an alternative method to measure body composition such as BIA or circumferences.

Assumptions and Sources of Error in BIA Assessment

The most basic assumptions of the BIA method are that the body's shape is a perfect cylinder and that body tissues will conduct electrical impulses along a path of least resistance (Heyward and Wagner 2004). Knowing this information, plus the length and cross-sectional area of the body, both electrical impedance and resistance can be measured and %BF calculated. Overall, these are acceptable assumptions.

The largest source of error in BIA measurement is the individual client (Heyward and Wagner 2004). As stated previously, it is important that the client follow the preassessment guidelines as closely as possible to ensure proper hydration status, and be free of anything (i.e., jewelry, metal devices, etc.) that can affect impedance to electrical flow through the body. A very minimal source of error is technician skill, especially compared to skinfold measurement. Most BIA analyzers are easy to prepare for use, with both proper electrode placement (traditional BIA) and skin preparation as the most important features of the preassessment process. Other factors that could be sources of error are choosing a valid BIA analyzer, and

the appropriate population-specific equations if calculating FFM to determine %BF (Houtkooper et al. 1996).

Normative Data and Interpretation

Table 3.2 can be used to compare %BF values that were determined using a variety of assessment methods; however, the data were generated from skinfold instrumentation and may be best suited for this method. You will use this normative chart for this lab and the next lab, which involves the assessment of body composition utilizing laboratory methods. To use this chart, find your sex-specific section and age-specific row and then determine where your %BF value falls according to the ratings at the top of the chart.

While it is important to interpret the level of %BF from a health risk appraisal viewpoint, it is equally as important to know and evaluate the current level of FFM and then monitor changes across the lifespan. Many health issues related to aging and chronic disease can be strongly linked back to changes in FFM, especially decreases in skeletal muscle mass (i.e., sarcopenia). However, an issue to consider is that there is no current consensus recommendation on what constitutes a healthy or optimal range of body composition due to other related factors such as age, race or ethnicity, activity level, and so forth (ACSM 2021). Thus, when using this chart, carefully interpret individual body composition values with this in mind and incorporate this information in the larger context of a full health and fitness screening.

BODY COMPOSITION DATA SHEET

Name _____ Sex _____ Date _____

Age _____ Height _____ cm Weight _____ kg

Three-site skinfolds (values should be recorded in mm):

Men	Trial 1	Trial 2	Trial 3	Average	Women	Trial 1	Trial 2	Trial 3	Average
Chest					Triceps				
Abdomen					Suprailiac				
Thigh					Thigh				

Note: Trial 3 is done only if needed.

Sum of 3 skinfolds _____ mm

Three-site formula for men (Jackson and Pollock 1978):

Body density (g·cc^{-1}) = 1.10938 − 0.0008267 (sum of 3 skinfolds) + 0.0000016 (sum of 3 skinfolds)^2 − 0.0002547 (age)

Three-site formula for women (Jackson, Pollock, and Ward 1980):

Body density (g·cc^{-1}) = 1.0994921 − 0.0009929 (sum of 3 skinfolds) + 0.0000023 (sum of 3 skinfolds)^2 − 0.0001392 (age)

%BF _____ Fat mass _____ kg Fat-free mass _____ kg

Percentile rank and rating from table 3.2 _____

Seven-site skinfolds (values should be recorded in mm):

Site	Trial 1	Trial 2	Trial 3	Average	Site	Trial 1	Trial 2	Trial 3	Average
Chest					Midaxillary				
Abdomen					Suprailiac				
Thigh					Triceps				
Subscapular									

Note: Trial 3 is done only if needed.

Sum of 7 skinfolds _____ mm

Seven-site formula for men (Jackson and Pollock 1978):

Body density (g·cc^{-1}) = 1.112 − 0.00043499 (sum of 7 skinfolds) + 0.00000055 (sum of 7 skinfolds)^2 − 0.00028826 (age)

Seven-site formula for women (Jackson, Pollock, and Ward 1980):

Body density (g·cc^{-1}) = 1.097 − 0.00046971 (sum of 7 skinfolds) + 0.00000056 (sum of 7 skinfolds)^2 − 0.00012828 (age)

%BF _____ Fat mass _____ kg Fat-free mass _____ kg

Percentile rank and rating from table 3.2 _____

Additional skinfold sites (values should be recorded in mm):

Site	Trial 1	Trial 2	Trial 3	Average	Site	Trial 1	Trial 2	Trial 3	Average
Biceps					Calf				

Note: Trial 3 is done only if needed.

BIA measurement:

%BF _____   Fat mass _____ kg Fat-free mass _____ kg

BIA measurement 2 (if applicable):

%BF _____   Fat mass _____ kg Fat-free mass _____ kg

Percentile rank and rating from table 3.2 _____

Assessment of Body Composition: Laboratory Methods

Purpose of the Lab

When available, densitometry and dual-energy X-ray absorptiometry (DEXA) techniques can be used as reference methods to obtain body density (Db) and the other body composition components. Densitometry techniques use either hydrostatic weighing (HW) or air displacement plethysmography (ADP) to estimate body volume and then apply the relationship between body mass, body volume, and Db to solve for Db (Db = Body Mass / Body Volume). As discussed in the previous chapter, the body contains protein, fat, mineral, and water components. Different laboratory body composition methods measure combinations of these four components to estimate composition. In this chapter, we will discuss methods for measuring body composition in both two-component

and three-component models. Densitometry techniques are the gold standard to obtain fat mass (FM) and fat-free mass (FFM) in the two-component model of body composition (Brožek et al. 1963; Siri 1961). DEXA is considered the reference method for obtaining bone mineral, FM, and FFM in the three-component model of body composition. It is important to note that despite the ability of DEXA to derive metrics for bone mineral density, it is not necessarily the most valid means for obtaining percent body fat (%BF).

Hydrostatic Weighing

Hydrostatic weighing uses Archimedes' principle to estimate body volume by calculating the buoyancy of a participant submerged in a large tank of water. In other words, the amount of weight that a participant loses when submerged in water is proportional to the amount of water displaced by their total body volume. The body mass in the Db equation is a participant's dry weight wearing a tight-fitting swimsuit. Total body volume is calculated by subtracting the average net underwater weight (UWW_{net}) from dry weight (Body $Mass_{dry}$) and dividing the difference by the water density (D_{water}). In the two-component model, the density of FM is assumed to be 0.901 $g \cdot cc^{-1}$ while the density of FFM is 1.100 $g \cdot cc^{-1}$. Compared to the water, FM is less dense and will cause a participant to float, while FFM is denser than water and will cause a participant to sink. Therefore, a participant who has relatively less FM will be heavier when weighed underwater compared to a participant with relatively more FM.

The weighing is typically performed using either a load cell system attached to a platform (figure 4.1) or a mechanical autopsy scale attached to a chair (figure 4.2). Despite the difference in price between the two scale systems (~$6,000 for load cell versus ~$450 for autopsy scale) there is no difference in their individual accuracy when performing HW measurements (Moon et al. 2011). We also must account for the total volume of air in the body when calculating total body volume: the air remaining in the lungs after maximal exhalation (residual lung volume) and the air in the gastrointestinal system (gastric volume). Neglecting to account for these volumes of air can be a major source of error in HW. The two-component model assumes that anything causing the individual to float is due to FM and therefore any volume of air not accounted for in the equation will be assumed to be FM. Residual lung volume (RV) measurement is completed by a separate procedure, most often by nitrogen washout or oxygen and helium dilution. When RV is measured properly, the accuracy of HW is 0.21 to 1.04 %BF compared to 1.98 to 3.7 %BF when RV is estimated (Morrow et al. 1986). Information and procedures for RV testing and estimation can be found in lab 12. Gastric volume (GV) cannot be directly measured so it is always assumed to be a constant 100 mL or 0.1 L in the equation. Altogether, Db can be calculated using the following equation:

$$Db = \frac{Body\ Mass_{dry}}{[(Body\ Mass_{dry} - UWW_{net})/D_{water}] + (RV + GV)}$$

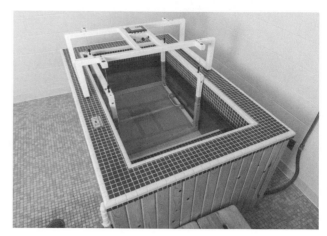

FIGURE 4.1   Load cell.

FIGURE 4.2   Autopsy scale.

Air Displacement Plethysmography

Another laboratory method used to evaluate FM and FFM is ADP, commonly known as the BOD POD (figure 4.3). Like HW, ADP uses displacement to estimate Db; however, it measures the displacement of air within an enclosed chamber rather than displaced water in an open tank. The BOD

POD collects a series of volume measurements in a two-chamber system and applies Poisson's law. Poisson's law states that volume (V) and pressure (P) are inversely related under adiabatic conditions, or $P_1/P_2 = (V_2/V_1)^\gamma$. Air in the chambers is under adiabatic conditions due to the air freely losing and gaining heat because it is compressed and expanded during the measurement process. The γ in Poisson's law is a constant equal to 1.4. Unfortunately, not all the air in the chamber is under adiabatic conditions. Small portions of isothermal air trapped in the clothing, near the skin, and in the lungs will compress more than adiabatic air by 40%. The BOD POD corrects for isothermal air near the skin's surface by using a participant's height and weight to calculate body surface area (Fields, Goran, and McCrory 2002). The amount of air in the lungs during normal tidal breathing (thoracic gas volume) is also measured while a participant is in the chamber through a sequence of breathing techniques into a small tube. Isothermal air in the clothing is minimized by the participant wearing compression-type clothing and a swim cap. Total body volume is measured as a participant displaces air when sitting inside the chamber. The volume of air is measured in the empty enclosed chamber and then again with a participant sitting in the chamber. During each measurement, a diaphragm that separates the chambers oscillates and produces small volume perturbations equal in magnitude and opposite in sign. These perturbations cause small pressure changes and are measured by the BOD POD. The difference in volume between the two conditions is calculated to represent body volume and then corrected by removing the isothermal air volumes (Fields, Goran, and McCrory 2002). Body density is then derived by dividing the individual's body mass by the body volume given by the BOD POD.

The BOD POD is an attractive means to assess body composition for individuals who have a difficult time submerging themselves underwater due to fear or discomfort. Furthermore, the BOD POD assessment requires less technical skill and time to administer compared to HW. Multiple studies comparing BOD POD to HW in adults and children have shown it to be a valid alternative. Mean group differences between BOD POD and HW range from −4.0 to 1.9 %BF in adults and −2.9 to 2.6 %BF in children, respectively (Fields, Goran, and McCrory 2002).

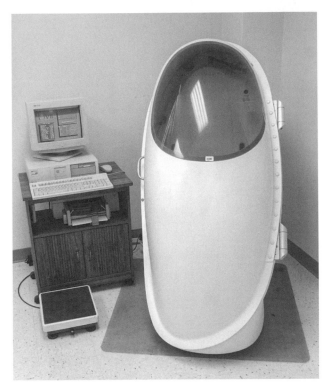

FIGURE 4.3   BOD POD.

Dual-Energy X-Ray Absorptiometry

While HW may be considered the gold standard of the two-component model of body composition, DEXA can be thought of as the research standard due to the ability to examine bone mineral, FM, and lean tissue (figure 4.4). Each of these components can be evaluated at the segmental level in addition to the whole body estimate. This allows researchers and practitioners to examine the status of and changes to specific areas of interest such as visceral FM, appendicular lean mass, and femoral neck and lumbar spine bone mineral density. The DEXA scan can differentiate between tissue types because bone, muscle, and fat have various levels of attenuation. In other words, the tissue properties of bone, muscle, and fat cause them to absorb or deflect X-ray beams differently from one other. The DEXA software analyzes attenuation by comparing the scan image pixels to those established from the reference material scanned during the calibration process. The attenuation of bone tissue is substantially greater than that of muscle and fat tissue and can be easily distinguished. Muscle and fat tissue

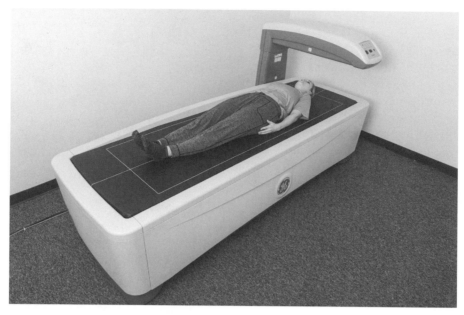

FIGURE 4.4   DEXA.

attenuation coefficients are also different from one another and can be quantified as well (Kohrt 1997).

Participants may also find the measurement process during DEXA to be more enjoyable than both HW and BOD POD because the procedure is performed in an open-air environment while the participant remains motionless in a supine position. Despite the utility of DEXA, research on the validity of DEXA to measure FM and FFM has shown it to be questionable compared to HW. A review by Kohrt (1997) summarized the difference in %BF between DEXA and HW ranges to be between −8.6% and 5.9% across 20 studies using males and females within a mean age range of 22 to 68 yr. Even so, using DEXA to examine status and changes in bone mineral density over the lifespan is a worthwhile endeavor. Low bone mineral content (osteopenia) currently affects 34 million Americans. Furthermore, the incidence of osteopenia in Americans 65 yrs of age and older is expected to grow from 13% to over 20% between 2010 and 2030. The main consequence of osteopenia is the effect of low bone mineral density on risk for fragility fractures, often as the result of a fall. Nearly 2 million fragility fractures occur annually in the United States and they have a substantial cost on the healthcare system and quality of life. Total direct and indirect costs of these fractures are expected to exceed $25 billion in the United States by 2025. In addition, a hip fracture is associated with a 1 yr mortality rate of ~20% (Burge et al. 2007; Varacallo et al. 2013).

Necessary Equipment or Materials

Hydrostatic Weighing

- Body weight scale
- Hydrostatic weighing tank
- Load cell weighing system and platform, if preferred
- Autopsy scale and chair, if preferred
- Tare weights
- Calibration weights (if using load cell system)
- Water thermometer
- Stadiometer (if estimating RV)

Air Displacement Plethysmography

- BOD POD system
- Calibration weights
- Calibration cylinder
- Body weight scale
- Stadiometer

Calibration weights, calibration cylinder, and body weight scale are part of the BOD POD system.

Dual-Energy X-Ray Absorptiometry

- Body weight scale
- Stadiometer
- DEXA scanner (General Electric, Hologic, or Norland)

Calibration of Equipment

Hydrostatic Weighing

- If using a load cell system, tare the scale with nothing on the platform. Place 4 kg on the platform and compare to the measured value. Calibrate if the measured value is not exactly 4 kg.
- Check the accuracy of the autopsy scale by hanging 4 kg from the scale. If the scale does not read exactly 4 kg, adjust the needle to 4 kg by turning the screw on the bottom of the scale (may vary based on model).

Air Displacement Plethysmography

- Follow the calibration procedure instructions in the BOD POD software system.

Dual-Energy X-Ray Absorptiometry

- Follow the calibration procedure using the calibration block in accordance with the system software instructions.

Procedures

Hydrostatic Weighing

Pretest Instructions

1. Participant should wear a tight-fitting swimsuit or one that will not trap air underwater. Jewelry should be removed.
2. Participant should bring a towel and any soap and shampoo desired for use after testing.

3. Participant should avoid eating foods that cause excessive gastrointestinal distress for at least 12 h prior to testing.
4. Participant should arrive normally hydrated and avoid physical activity 4 h or more prior to testing.
5. Participant should void completely prior to testing.

Testing

1. Weigh participant (dry) in their swimsuit to the nearest 0.1 kg.
2. Instruct the participant to shower without using any soap or shampoo.
3. Measure and record tare weight of chair, water temperature, and water density (see table 4.1) before participant enters tank. If using a load cell system, zero the scale with the tank completely empty and perform the calibration process using calibration weights.
4. Participant can enter the tank. Instruct the participant to quickly submerge their head underwater and remove air bubbles from hair. Participant should then run their hands over their entire body, removing air bubbles trapped against skin and in the swimsuit. If using a load cell system, zero the scale with the participant in the water but not touching the platform.
5. Participant should set up in measurement position, seated toward the front edge of the chair if using the autopsy scale system or sitting cross-legged on the platform in the load cell system.
6. Instruct participant on the entire procedure and what to expect during the measurement process.
7. The participant should begin exhaling to near full exhalation and then slowly lean forward until their head and body are fully submerged. The participant should continue to exhale until no air bubbles are observed. The participant should be instructed to remain as still as possible while underwater.
8. Once all the air is expelled, the measurement is taken. If using an autopsy scale system, record the highest stable weight. If using a load cell system, verbalize a 3 s count after

TABLE 4.1 Relative Density of Water at Various Temperatures

Temperature (°C)	Temperature (°F)	Density (g·cc⁻¹)
30	86	0.9957
31	88	0.9954
32	90	0.9951
33	91	0.9947
34	93	0.9944
35	95	0.9941
36	97	0.9937
37	99	0.9934
38	100	0.9930
39	102	0.9926
40	104	0.9922

air bubbles disappear. In both cases, knock on the tank to instruct the participant to emerge from the water. Record the weight to the nearest 0.01 kg.

9. If using a load cell system, select the recommended samples to determine an underwater weight in accordance with the manufacturer guidelines. Often, 2 to 3 s provides enough samples to collect an accurate weight.

10. The procedure should be repeated until three trials are completed that are within 0.1 kg of one another.

11. Average these trials and subtract the tare weight to get the UWW_{net}.

12. Using the following equation, calculate Db:

$$Db = \frac{\text{Body Mass}_{dry}}{[(\text{Body Mass}_{dry} - UWW_{net})/D_{water}] + (RV + GV)}$$

13. Convert Db to %BF by using the appropriate population-specific conversion equation in lab 3.

14. Calculate FM and FFM using the following equations:

$$FM = \text{Body Mass}_{dry} \times \%BF$$

$$FFM = \text{Body Mass}_{dry} - FM$$

Example: Body Mass$_{dry}$ = 75 kg, %BF = 28.1%

$$FM = 75 \text{ kg} \times 0.281 = 21.1 \text{ kg}$$

$$FFM = 75 \text{ kg} - 21.1 \text{ kg} = 53.9 \text{ kg}$$

15. Interpret %BF using table 3.2.

Air Displacement Plethysmography

Pretest Instructions

1. Participant should wear compression-type clothing and a swim camp. Jewelry should also be removed.

2. Participant should avoid eating, drinking, and exercise 2 h prior to testing.

3. Participant should void completely just prior to testing.

Testing

1. Ensure that the BOD POD system is turned on and warmed up for at least 30 min prior to calibration.

2. If measuring thoracic gas volume, install filter and breathing tube in the chamber.

3. Following the on-screen BOD POD system instructions, perform the weight scale and chamber calibration process.

4. Measure and record the participant's height (cm) using a stadiometer and weight using the BOD POD.

5. The participant can enter the chamber and sit with their back against the wall. Instruct the participant to remain still and breathe normally during the measurement procedure.

6. Follow the on-screen BOD POD instructions during the body volume measurement process.

7. Prior to thoracic gas volume measurement, instruct the participant on breathing technique during the procedure.

8. Ensure that the participant can view the breathing cues on the system screen during the thoracic gas volume measurement process.

9. Upon acceptable agreement and successful completion of the thoracic gas volume measurement, the participant can exit the BOD POD chamber.

10. Print the BOD POD results for body composition analysis and interpretation.

11. Interpret %BF using table 3.2.

Dual-Energy X-Ray Absorptiometry

Pretest Instructions

1. Participant should wear compression-type clothing or dress down to a single layer of clothing. Jewelry should also be removed.

2. Participant should void completely just prior to testing.

Testing

1. Perform calibration process in accordance with the manufacturer guidelines.

2. Measure the participant's height (cm) and weight (kg).

3. Instruct the participant to lie down in a supine position on the scanning bed with their head at the appropriate end of the machine.

4. Follow on-screen instructions for participant positioning and throughout the test; instructions will depend on the type of scan being administered (e.g., whole body, lumbar spine, femoral neck).

5. The participant should remain still and quiet throughout the measurement process.

6. Ensure that quality assurance has been met. If quality assurance is not met, adjust participant positioning as needed.

7. Print DEXA results for body composition analysis and interpretation.

8. Interpret %BF using table 3.2.

Common Errors and Assumptions

Common Errors for Hydrostatic Weighing

- The participant does not fully exhale air while submerged.

- If using an autopsy scale system, excess movement underwater can make it difficult to get an accurate reading. It helps to steady the scale with your hand, but remember to remove your hand while reading the weight. With either weighing system, instruct the participant to make all submersion movements slow and subtle in order to minimize water turbulence.

- When calculating Db, predicted RV is used rather than measured RV.

- Appropriate unit precision is not used when measuring Body Mass$_{dry}$, UWW, or RV.

- Rounding Db. Always carry calculated Db out to five decimal places before converting to %BF.

Assumptions in the Two-Component Model of Body Composition

- The density of the FM is 0.901 g·cc^{-1}.
- The density of the FFM is 1.100 g·cc^{-1}.
- The densities of FM and FFM are the same for all individuals.
- The densities of the components of the FFM (water, protein, bone mineral) are constant within an individual, and their proportional contribution to the lean component remains constant.
- The individual being measured differs from the reference body (research) only in the amount of fat. The FFM of the reference body is assumed to be 73.8% water, 19.4% protein, and 6.8% bone mineral.

BODY COMPOSITION: LAB METHODS WORKSHEET

Name _____ Sex _____ Date _____

Age _____   Water temperature _____°C

Height _____ cm   Water density (D_{water}) _____ g·cc^{-1}

Body Mass$_{dry}$ _____ kg   RV avg _____ L

Hydrostatic Weighing Trials

1. _____ kg   6. _____ kg   Average UWW _____ kg

2. _____ kg   7. _____ kg   Tare weight _____ kg

3. _____ kg   8. _____ kg   UWW$_{net}$ _____ kg

4. _____ kg   9. _____ kg   Db _____ g·cc^{-1}

5. _____ kg   10. _____ kg

$$Db = \frac{\text{Body Mass}_{dry}}{[(\text{Body Mass}_{dry} - \text{UWW}_{net})/D_{water}] + (RV + GV)}$$

%BF _____   Fat weight _____ kg   Fat-free weight _____ kg

BOD POD

Lung volume _____   %BF _____

Body volume _____   Fat weight _____ kg

Body density _____   Fat-free weight _____ kg

Evaluation

_____

_____

_____

_____

_____

_____

_____

_____

_____

_____

_____

Assessment of Resting Metabolic Rate and Weight Management Strategies

Purpose of the Lab

We have discussed many ways to assess and quantify overweight and obesity in previous labs. From using height and weight to calculate body mass index (BMI) to measuring the thickness of subcutaneous fat deposits at specific landmarks to weighing an individual underwater to estimate body volume, the goal has been to calculate fat mass (FM) and fat-free mass (FFM). As mentioned in lab 2, BMI can be used to classify individuals as overweight (25.0-29.9 kg/m^2), class I obese (30.0-34.9 kg/m^2), class II obese (35.0-39.9 kg/m^2), or class III obese ($\geq 40.0\ kg/m^2$) (World Health Organization 2000). The inference when using BMI is that excess weight within individuals across overweight and obese categories is specific to FM, not FFM. Therefore, estimations of body composition (field or laboratory) are preferred over BMI when examining overweight and obesity on an individual level.

The World Health Organization (2021) estimates that globally over 1.9 billion adults aged 18 yr or older were overweight and 650 million adults were obese in 2016. These numbers represent 39% and 13% of the worldwide population, respectively. Moreover, these numbers tripled from 1975 to 2016. These trends are not unique to any country, age group, or socioeconomic class; it is an issue that affects humankind. Only 4% of children between the ages of 5 and 19 were overweight in 1975 but that number grew to 18% in 2016. Similarly, the prevalence of obesity in children grew from 1% to 7% in that same time frame. The burden of overweight and obesity is felt by healthcare systems across the world. For example, the annual direct medical cost of an overweight individual is $266 higher and the cost of an obese individual is $3,748 higher than a normal weight individual in the United States (Tsai, Williamson, and Glick 2011; Biener, Cawley, and Meyerhoefer 2018). It is important to note that these numbers reflect the direct medical cost of overweight and obesity, which is the cost to directly treat the accompanying health complications associated with them. They do not factor in the indirect costs of overweight and obesity such as increased absences from work due to illness or appointments, or the reduction in productivity while at work. It has been reported that the indirect cost of obesity is upward of 2.5 times greater than that of the direct medical cost (Levine et al. 2019).

Excess body fat is associated with a myriad of health complications. Overweight and obesity have been associated with an increased prevalence of hypertension (HTN). Males and females who are overweight are 2.1 times and 1.9 times more likely to develop HTN than individuals of normal weight, respectively. Moreover, 38% of men who are obese and 32% of women who are obese have HTN (Khaodhiar, McCowen, and Blackburn 1999). With every 1 kg increase in excess body weight there is a 5% increase in risk for HTN, but weight loss can combat this trend; losses of 2 to 4 kg can reduce systolic blood pressure by 3 to 8 mmHg (Harsha and Bray 2008).

Obesity is also a major risk factor for the development of type 2 diabetes mellitus (T2DM). Excess abdominal obesity by way of visceral fat accumulation appears to be the most significant because it disrupts endocrine homeostasis and increases the risk for insulin resistance, one of the underlying mechanisms in T2DM. It is important to note that 85.2% of individuals with T2DM are overweight or obese (CDC 2004). Both HTN and T2DM are major risk factors for coronary artery disease (CAD) development and overall mortality, but there is some evidence to suggest that obesity may be an independent risk factor for CAD development (Jahangir, De Schutter, and Lavie 2014). Conversely, an "obesity paradox" exists showing a reduction in overall cardiovascular disease (CVD) mortality once an individual who is overweight or obese has developed CVD. Individuals who are overweight or obese and have CAD have similar or even decreased adverse health outcomes compared to individuals who are normal weight with CAD. Numerous mechanisms are purported to explain the obesity paradox; however, it is generally accepted that lean body mass preservation and

cardiorespiratory fitness level are better predictors of CAD progression and mortality than BMI or adiposity. Regardless of the questions surrounding obesity and CAD, weight loss remains an effective way to reduce CAD risk factor development and progression (Jahangir, De Schutter, and Lavie 2014). Excess body fat may cause alterations in endogenous hormone metabolism, which can disrupt the regular process of cell balance within the body and lead to increased risk of cancer development, namely colon, breast, endometrial, kidney, esophageal, and gastrointestinal cancers (Bianchini, Kaaks, and Vainio 2002).

Developing and implementing weight management programs can be a complex undertaking that requires allied health professionals across various scopes of practice working together to ensure successful client outcomes. These programs may focus on weight gain, loss, or maintenance, but we will highlight weight loss in this chapter due to the prevalence of overweight and obesity across the globe. A weight management program is centered around the concept of energy balance, the difference between the amount of energy consumed and expended each day. This energy is quantified in kilocalories (kcal). By definition, a kcal is the amount of heat needed to increase the temperature of 1 kg of water 1 °C. Foods have different amounts of kcal within them based on their composition of fats, proteins, and carbohydrates. Eating food containing any amount of fat, protein, or carbohydrate adds to the energy consumption portion of the energy balance scale. Conversely, there are four factors that contribute to the energy expenditure end of the energy balance scale. These factors are resting metabolic rate (RMR), nonexercise activity thermogenesis (NEAT), thermic effect of food (TEF), and exercise activity thermogenesis (EAT). RMR is synonymous with the concept of basal metabolic rate (BMR), only differing in the way that resting energy expenditure is measured. BMR measurements are taken in a tightly controlled setting over the course of 10 to 24 h, while an entire RMR assessment may only take 30 to 60 min. The values obtained in the carefully selected 10 min data set are then extrapolated out to represent RMR over the course of a day. RMR represents the amount of energy needed to sustain physiological function of the body's organ systems at rest and comprises ~60% to 70% of an individual's total daily energy expenditure

(TDEE). The energy required to sustain the organ system at rest depends on many factors including the following (Ruggiero and Ferrucci 2006):

- An individual's RMR tends to decline throughout the entire lifespan, and men tend to have higher RMR than women due to differences in body composition and hormonal patterns.

- Hormones such as testosterone, growth hormone, and insulin-like growth factor-1 will upregulate RMR.

- RMR is proportional to body size, where individuals with greater body surface area will have a greater RMR.

- Individual body composition has a substantial impact on RMR.

- Lean muscle mass, or more specifically the fat-mass-to-lean-mass ratio, increases RMR.

- Physical activity will slightly increase RMR due to elevated oxygen consumption after cessation of an exercise session.

- Finally, RMR is greater in the presence of chronic diseases such as chronic obstructive pulmonary disease and congestive heart failure. The increase in RMR is one of the main contributors to the loss in muscle mass typically observed in many chronic diseases.

The thermic effect of food is the energy required to aid the digestive system in the breakdown of food and makes up ~8% to 15% of TDEE. While RMR and TEF are a relatively constant percentage of TDEE, NEAT and EAT are much more variable. Nonexercise activity thermogenesis is all unplanned movements completed over the course of a day such as occupational tasks, fidgeting, sitting/standing, talking, or ambulation. Due to the variability between individuals in these daily movement tasks, NEAT can be ~15% of TDEE in individuals who are sedentary to ~50% of TDEE in individuals who are highly active. Exercise activity thermogenesis is also activity-specific energy expenditure except that it is planned, structured, and purposeful physical activity for durations of 10 min or more. EAT can represent anywhere between ~15% to 30% of TDEE (Aragon et al. 2017). Given that RMR represents the largest and most consistent portion of TDEE, we measure and apply it when designing an individualized weight management program.

As mentioned earlier, this chapter will focus on weight loss programming. The simplest approach to a weight loss program is to place the client in a caloric deficit each day, and there are guidelines to ensure safe and effective negative caloric balance. Setting a target weight goal is important for client motivation and program tracking, but practitioners should also focus on the development of healthy lifelong behaviors such as regular exercise and healthy eating habits. A weight loss program should center around a well-balanced diet that meets all recommended nutritional requirements. The nutrition plan should include intake of at least 1,200 kcal·day^{-1} and place the client in a total caloric deficit of no more than 1,000 kcal·day^{-1}. Despite the allure of many commercially available weight loss programs, weight loss should be a gradual 1 to 2 lb (0.45-0.91 kg) loss per week. In addition, all the weight lost should be fat mass, with an emphasis on maintaining lean mass. Proper nutrition and exercise strategies should aim to mitigate loss in lean muscle mass. For example, if we adhere to the guideline of a deficit of no more than 1,000 kcal·day^{-1} then it would take approximately 3.5 d to lose 1 lb of fat (1 lb fat = 3,500 kcal). With these general principles in mind, we can begin the steps to design an individual weight loss program.

Step 1: Review risk stratification and measure body composition.

See lab 1 for preparticipation screening information. It is important when starting a weight loss program to complete a health history questionnaire to identify comorbidities, orthopedic and musculoskeletal issues, or special considerations regarding overall fitness capacity. We will also need to measure body composition to set a weight goal. Review labs 3 and 4 to select the most appropriate body composition assessment.

Here is an example: Your client is a 42-year-old female high school science teacher. They are 5 ft 4 in. (162.6 cm), 164 lb (74.5 kg), and 37 %BF. Their health history questionnaire showed no presence of chronic disease or comorbidities. Given the range for a healthy recommended %BF for a female your client's age, you determine that 32 %BF is an appropriate and realistic goal for an 8 wk weight loss program. To calculate their end weight goal, use the following series of calculations:

Fat Mass (FM) = Body Weight × %BF

Fat-Free Mass (FFM) = Body Weight − FM

Goal Weight = FFM / (1 − Goal %BF)

FM = 164 lb × 0.37 = **60.7 lb (27.5 kg)**

FFM = 164 lb − 60.7 lb = **103.3 lb (46.9 kg)**

Goal Weight = 103.3 lb / (1 − 0.32)
= 103.3 lb / (0.68) = **152 lb (68.9 kg)**

The calculated goal is to lose 12 lb (5.4 kg) of fat in 8 wk. This agrees with the weight loss program guideline of 1 to 2 lb (0.45-0.91 kg) of weight loss per week.

Step 2: Determine dietary intake.

The ideal scenario is to work closely with a registered dietician or other trained dietetics professional. The client should keep a food record for anywhere between 3 to 7 d, logging details for all food consumption during that time. Keep in mind that the prevalence of underreporting caloric intake on food logs can range between 18% and 54% (Macdiarmid and Blundell 1998). There are numerous food logging programs available online or on smartphone apps if access to a dietician is limited. The average daily caloric intake will give the dietician a starting point for the energy consumption portion of the energy balance equation.

Step 3: Estimate or measure RMR.

The primary lab experience in this chapter is the measurement of RMR via indirect calorimetry. The equipment, setup, and procedures for RMR testing will be discussed in later sections. A quick and inexpensive alternative to measuring RMR is using a predictive equation. The two most widely used equations are the Harris-Benedict and Mifflin et al. equations. Both prediction equations are sex-specific equations that require body mass (kg), height (cm), and age (years) and calculate RMR in $kcal \cdot day^{-1}$. The equations are listed as follows, where BM = body mass and ht = height.

Harris-Benedict equation (Harris and Benedict 1918)

Men: RMR = 66.473 + 13.751(BM) + 5.0033(ht) − 6.755(age)

Women: RMR = 655.0955 + 9.463(BM) + 1.8496(ht) − 4.6756(age)

Mifflin et al. equation (Mifflin et al. 1990)

Men: RMR = 9.99(BM) + 6.25(ht) − 4.92(age) + 5.0

Women: RMR = 9.99(BM) + 6.25(ht) − 4.92(age) − 161

Example:

Estimated RMR = 9.99(74.5 kg) + 6.25(162.6 cm) − 4.92(42 yr) − 161 = **1,393 kcal·day⁻¹**

Measured RMR = **1,472 kcal·day⁻¹**

Step 4: Apply the correction factor for occupational activity level.

In addition to RMR, the level of activity throughout the day will be a major contributing factor to TDEE. Occupations that have greater demands for movement and physical exertion will expend more calories compared to those that require little movement. For example, an individual who paints the interior of homes would expend more calories throughout a working day than a computer programmer spending most of the day sitting. The goal of this step is to match the activity level required in the example occupations to the occupational demands of the client. The activity levels and example occupations are as follows:

Sedentary: Inactive

Lightly active: Most professionals, such as teachers, homemakers, office professionals, or shop workers

Moderately active: Farm workers, active students, nonactive military service, or light industry workers

Very active: Forestry workers, professional athletes, unskilled labor, or active-duty military service

Extremely active: Construction workers, lumberjacks, or blacksmiths

Once the occupational activity level (OAL) has been identified, apply the additional energy expenditure required for each level of activity based on sex. For example, a female high school science teacher has an occupational metabolic demand 35% greater than RMR alone.

Sedentary: Men = 15%; Women = 15%

Lightly active: Men = 40%; Women = 35%

Moderately active: Men = 50%; Women = 45%

Very active: Men = 85%; Women = 70%

Extremely active: Men = 110%; Women = 100%

Example:

OAL = RMR × OAL correction factor

OAL = 1,472 kcal·day^{-1} × 0.35 = **515 kcal·day^{-1}**

Energy Expenditure = 1,472 kcal·day^{-1}
+ 515 kcal·day^{-1} = **1,987 kcal·day^{-1}**

Step 5: Set a caloric deficit goal.

Remember, an effective weight loss program should aim for a caloric deficit of ≤1,000 kcal·day^{-1} This deficit should be accomplished through a combination of restricted caloric intake and increased exercise energy expenditure. Although caloric restriction in the diet can lead to positive weight loss outcomes, the evidence suggests that combining exercise with moderate diet restriction (~500-700 kcal·day^{-1}) leads to more significant weight loss than diet restrictions alone (Donnelly et al. 2009). Exercise energy expenditure in the 1,200 to 2,000 kcal·wk^{-1} range is not only sufficient to prevent weight gain in adults but is also adequate for moderate weight loss. Expending 1,500 kcal·wk^{-1} is more manageable when accomplished over 4 to 5 d per week than 2 to 3 d per week. Exercise volume (days and duration) is the main exercise prescription principle to focus on during weight loss programming. The exercise days should emphasize aerobic exercise for caloric expenditure but should also include resistance training for the purpose of muscle mass preservation.

Example:

Deficit goal = 1.5 lb of fat loss per week

1.5 lb of fat loss/wk × 3,500 kcal·lb^{-1} of fat
= **5,250 kcal·wk^{-1}**

Exercise deficit: 5 d/wk and expend 300 kcal per exercise session

5 d/wk × 300 kcal·day^{-1} = **1,500 kcal·wk^{-1}**

Nutrition deficit: 5,250 kcal·wk^{-1}
− 1,500 exercise kcal·wk^{-1} = **3,750 kcal·wk^{-1}**

3,750 kcal/7 d = **536 kcal·day^{-1} deficit**

Your client should plan on restricting 536 kcal·day^{-1} throughout the 8 wk weight loss program. On exercise days, your client will be in a total deficit of 836 kcal (536 kcal through diet + 300 kcal through exercise). Both exercise and non-exercise days align with our 500 to 1,000 kcal·day^{-1} deficit recommendation for healthy weight loss. A detailed nutritional plan should be provided by a dietary professional to achieve the 536 kcal·day^{-1} caloric restriction.

Necessary Equipment or Materials

- Metabolic analyzer
- Body weight scale
- Stadiometer
- Treatment table
- Pillow
- Light blanket (if needed)
- Fan
- All measurements should be performed in a quiet, dimly lit or dark room.
- A room temperature range of 20 to 25 °C (68-77 °F) is recommended.

Calibration of Equipment

Calibration of the metabolic analyzer should be performed in accordance with the manufacturer guidelines. Generally, indirect calorimetry systems need to warm up for at least 30 min prior to calibration. They are then calibrated through a series of gas and flow meter calibration steps.

Procedures

Pretest Instructions

Your client should do the following:

1. Fast for at least 5 h prior to testing.
2. Refrain from alcohol and nicotine for at least 2 h and caffeine for at least 4 h prior to testing.
3. Avoid moderate physical activity for at least 2 h and vigorous physical activity for at least 14 h prior to testing.

Testing

1. Upon arrival, obtain client's height and weight.
2. Enter all client information into the RMR measurement program.
3. Position the client in a supine position on the treatment table. A rest period of 10 to 20 min is recommended prior to beginning the RMR procedure.
4. Place a pillow under the client's head and a foam wedge or pillow under the knees for support. A blanket should be supplied to keep the client in a comfortable temperature range.
5. Turn on and position a fan in the room to ensure proper room air circulation.
6. Place the measurement hood/mask on the client in accordance with the manufacturer guidelines for RMR measurement.
7. Instruct the client to remain calm and still during the entire measurement process. They should relax but refrain from falling asleep.

8. Ensure that the testing room is as quiet and dark as possible, and exit the room.
9. Return every 5 to 10 min to verify the quality of measured RMR data. If applicable, make any needed adjustments to the metabolic system in accordance with manufacturer guidelines.
10. After 20 to 30 min of testing have elapsed, remove the testing hood/mask from the client.
11. Discard the first 5 min of testing data. Select a 5 min portion of testing data that achieves a coefficient of variation $\leq 10\%$ for oxygen ($\dot{V}O_2$) consumption and carbon dioxide (CO_2) production.
12. Verify a respiratory exchange ratio (RER) between 0.7 and 1.0 for additional quality assurance.

Following the procedures described, complete a RMR assessment. Compare your RMR with your estimated RMR using the lab 5 worksheet.

Case Study for RMR and Weight Management

For the case study, please answer the questions that follow each example. You will need to rely on information from labs 1 and 3 to complete the case study.

Demography

Age: 41

Height: 5 ft 11 in. (180.3 cm)

Weight: 235.4 lb (107 kg)

Sex: Male

Race or Ethnicity: Black or African American

Family History

Your client's family history reveals the father (age 67) was recently diagnosed with CAD and had stents placed in the left anterior descending and circumflex coronary arteries. The client's grandfather (deceased at age 85) had aortic valve disease with a bioprosthetic valve replacement (diagnosed at age 70, replaced at age 75).

Medical History

Present Conditions

A recent physical assessment revealed the following health information. Your client's resting heart rate (HR) was 84 beats·min^{-1} and resting blood pressure (BP) was 128/78 mmHg (confirmed from a previous measurement). The blood chemistry panel showed a total cholesterol of 192 mg·dL^{-1}, a high-density lipoprotein (HDL) of 37 mg·dL^{-1}, a low-density lipoprotein (LDL) of 151 mg·dL^{-1}, and a triglyceride (TGL) level of 118 mg·dL^{-1}. Fasting blood glucose (FBG) was measured at 97 mg·dL^{-1}. A Rockport 1-mile walk test revealed a predicted maximal oxygen consumption ($\dot{V}O_2$max) value of 39.1 mL·kg^{-1}·min^{-1}. No other data are currently available.

Past Conditions

Your client does not report any past medical issues.

Behavior and Risk Assessment

The client works as a shift manager at a local retail store. The client presents to your community-based fitness program with a weight loss goal in mind. Your client's healthcare provider prefers trying lifestyle modification to reduce excess weight before considering medication. Your client walks their dog for 20 to 30 min most days of the week. They currently do not engage in any resistance training. A 3 d dietary recall analysis shows a 2,800 to 3,500 calorie per day diet. Average skinfold measurements were as follows: chest: 22 mm; abdomen: 34 mm; thigh: 27 mm.

1. What are their cardiovascular risk factors?
2. Calculate %BF, FM, and FFM from skinfold measurements.
3. Estimate RMR by using the average of the two RMR prediction equations.
4. Apply the OAL correction factor.
5. Set a caloric deficit goal.

> continued

Case Study for RMR and Weight Management   > continued

CASE STUDY ANSWER KEY

Cardiovascular risk factors	BMI $\geq$30 kg·m^{-1}, HDL < 40 mg·dL^{-1}
%BF, FM, and FFM from skinfold measurements	Three-site equation (Jackson and Pollock 1978): chest: 22 mm; abdomen: 34 mm; thigh: 27 mm Body density (g·cc^{-1}) = 1.10938 − 0.0008267 (83) + 0.0000016 (83)^2 − 0.0002547 (41) Body density (g·cc^{-1}) = **1.04134 g·cc^{-1}** %BF = (486/Db) − 439 = (486 / 1.04134) − 439 = **27.7 %BF** Goal %BF = 20 %BF FM = 235.4 lb × 0.277 = **65.2 lb (29.6 kg)** FFM = 235.4 lb − 65.2 lb = **170.2 lb (77.2 kg)** Goal weight = 170.2 lb / (1 − 0.20) = 170.2 lb / (0.80) = **213 lb (96.6 kg)**
Estimated RMR using the average of the two RMR prediction equations	Men: RMR = 66.473 + 13.751(107) + 5.0033(180.3) − 6.755(41) = **2,163 kcal·day^{-1}** Men: RMR = 9.99(107) + 6.25(180.3) − 4.92(41) + 5.0 = **1,999 kcal·day^{-1}** Average RMR = **2,081 kcal·day^{-1}**
OAL correction factor applied	OAL = RMR × OAL correction factor OAL = 2,081 kcal·day^{-1} × 0.40 = **832 kcal·day^{-1}** Energy expenditure = 2,081 kcal·day^{-1} + 832 kcal·day^{-1} = **2,913 kcal·day^{-1}**
Caloric deficit goal	Total weight loss goal = 235 lb − 213 lb = 22 lb (10 kg) Deficit goal = 1.75 lb (0.79 kg) of fat loss per week for 12 wk 1.75 lb of fat loss per wk × 3,500 kcal·lb^{-1} of fat = **6,125 kcal·wk^{-1}** Exercise deficit: 5 d/wk and expend **350 kcal per exercise session** 5 d/wk × 350 kcal·day^{-1} = **1,750 kcal·wk^{-1}** Nutrition deficit: 6,125 kcal·wk^{-1} − 1,750 exercise kcal·wk^{-1} = **4,375 kcal·wk^{-1}** 4,375 kcal / 7 d = **625 kcal·day^{-1} deficit** Our client should plan on restricting 625 kcal·day^{-1} throughout the 12 wk weight loss program. On exercise days, they will be in a total deficit of 975 kcal (625 kcal through diet + 350 kcal through exercise). Both exercise and nonexercise days align with our 500-1,000 kcal·day^{-1} deficit recommendation for healthy weight loss. A detailed nutritional plan should be provided by a dietary professional to achieve the 625 kcal·day^{-1} caloric restriction.

RESTING METABOLIC RATE

Name _____ Date _____

Age _____   Height _____ cm   Weight _____ kg   ☐ Female ☐ Male

Occupational Activity Level _____

Part 1: Estimate RMR

Harris-Benedict equation (Harris and Benedict 1918)

Men: RMR = 66.473 + 13.751(BM) + 5.0033(ht) − 6.755(age)

Women: RMR = 655.0955 + 9.463(BM) + 1.8496(ht) − 4.6756(age)

Estimated RMR _____ kcal·day^{-1}

Mifflin et al. equation (Mifflin et al. 1990)

Men: RMR = 9.99(BM) + 6.25(ht) − 4.92(age) + 5.0

Women: RMR = 9.99(BM) + 6.25(ht) − 4.92(age) − 161

Estimated RMR _____ kcal·day^{-1}

Average estimated RMR _____ kcal·day^{-1}

Part 2: Complete an RMR Assessment

Measured RMR _____ kcal·day^{-1}

Part 3: Comment on the Difference Between Estimated and Measured RMR

_____

_____

_____

_____

_____

_____

_____

_____

_____

_____

_____

_____

_____

_____

From J. Janot and N. Beltz, *Laboratory Assessment and Exercise Prescription* (Champaign, IL: Human Kinetics, 2023).

Metabolic Calculations and Calculation of Energy Expenditure

Purpose of the Lab

It is generally accepted that oxygen consumption ($\dot{V}O_2$) is the best measure of cardiorespiratory fitness during maximal exercise and also the best estimate of exercise intensity and energy expenditure during exercise (Astrand et al. 2003; Beltz et al. 2016; Wagner et al. 2020). Thus, exercise professionals should evaluate $\dot{V}O_2$ during fitness assessments and utilize it when designing aerobic exercise programs whenever possible. We will see that the measurement of $\dot{V}O_2$ through indirect calorimetry during exercise requires specialized equipment and training (discussed later in lab 8) and can be difficult to apply in certain practical situations. Therefore, when it is impractical to measure $\dot{V}O_2$ in this manner, we must employ alternative methods to estimate $\dot{V}O_2$ during exercise.

One such method to reasonably estimate $\dot{V}O_2$ during steady-state, submaximal aerobic exercise is through the application of the American College of Sports Medicine (ACSM) metabolic equations (ACSM 2021). Although equations designed to predict the metabolic intensity of exercise have existed since the 1950s, the ACSM metabolic equations remain the most widely applied (Moore 2019). The metabolic equations were developed and refined over time so that exercise professionals could ascertain the total oxygen cost of common activities (e.g., walking, running, cycling, etc.) if certain variables such as speed, grade, or workload are known. If the oxygen cost of an activity is known, then the kcal or energy expenditure for that activity can be calculated as well. For example, this is especially helpful information for the exercise professional to utilize for exercise prescriptions that might involve a weight management component, or in clinical populations where safety and effectiveness of exercise is vital. Additionally, if the exercise professional has an idea of what $\dot{V}O_2$ or MET (metabolic equivalent of task) level they want to prescribe for their client, these equations can be used to solve for the workload (speed, grade, step rate, power output, etc.) that would be equivalent to the specific MET level or $\dot{V}O_2$. Applied in these ways, the metabolic equations have the flexibility to solve for either $\dot{V}O_2$ or workloads during exercise.

The purpose of this lab is to introduce you to the ACSM metabolic equations and show you how to apply them. In this lab, you will learn what variables are involved in each equation and the assumptions behind these equations. In addition, you will be provided with opportunities to solve practical problems using these equations and learn the steps for the calculation of energy expenditure. You will also be using these equations in a subsequent lab (lab 7) involving the prediction of $\dot{V}O_2$max from exercise tests and in the design of the aerobic exercise prescription. Therefore, it is crucial to develop a deep understanding of these concepts in this lab.

Basic Assumptions of the ACSM Metabolic Equations

Since the ACSM metabolic equations involve the prediction of $\dot{V}O_2$, there are recommendations that must be followed and assumptions to consider in order to maximize the predictive accuracy of the equations. The basic assumptions for these equations are as follows (ACSM 2021):

1. $\dot{V}O_2$ for a given work rate or workload is uniform: This assumption refers to the notion that oxygen cost for a given work rate of exercise is the same for all individuals. This is a good assumption if you were to compare multiple exercise bouts of the same work rate for the same person. However, it is not a good assumption if you are comparing $\dot{V}O_2$ values at similar work rates between individuals. The most influential factor is varying mechanical efficiency from one person to another.

2. Steady-state exercise is performed by the individual: These equations were developed by ACSM for the estimation of $\dot{V}O_2$ during submaximal exercise intensities; as a result, they should not be used during non-steady-state exercise such as near or at $\dot{V}O_2$max. This assumption is a key fea-

ture for the predictive accuracy of these equations, and steady-state exercise can be very easily established during exercise. To ensure that the client is exercising at a steady state, we would need to measure heart rate (HR) over a period of time (~3-5 min) at a constant workload. In general, if the HR after 3 to 5 min of exercise does not fluctuate by more than ±6 beats·min⁻¹ at a constant workload, then steady-state exercise has been achieved.

3. The effects of extraneous variables are minimized: There are many factors, such as heat, cold, altitude, gait, and movement efficiency, which can affect the oxygen cost of exercise beyond the actual workload. The best way to address some of these factors is to control the environment, such as exercising indoors. However, this may not be feasible for clients who desire to exercise outdoors. Therefore, this assumption falls within the acceptable error category when calculating for $\dot{V}O_2$.

4. The equipment used is calibrated: This is an easy assumption to meet with correct preparation of modalities prior to testing.

5. The factors (e.g., speed, step rates, work rates, etc.) used in these equations to estimate $\dot{V}O_2$ are all within recommended ranges: This is an easy assumption to meet with careful attention to detail when calculating for $\dot{V}O_2$. These recommended ranges are discussed at length in the next section.

The ACSM Metabolic Equations

The five ACSM metabolic equations that we will be using in this lab are listed here. The unit for all $\dot{V}O_2$ values calculated from these equations is expressed in mL·kg⁻¹·min⁻¹. The equations (with the exception of the arm ergometry equation) are organized into three separate components, as follows:

1. A resting component, which is set at a constant of 3.5 mL·kg⁻¹·min⁻¹ (1 MET) for all equations,

2. a horizontal component, which addresses the oxygen cost of moving our body mass forward horizontally, and

3. a vertical or resistance component, which addresses the oxygen cost of either moving

our body mass against gravity (walking, running, stepping) or moving against some form of external load or resistance (cycling and arm ergometry) (ACSM 2021).

Please pay close attention to the units for each variable and the recommended ranges to be used for speed, step rates, and work rates when using these equations.

Walking

The ACSM walking equation (horizontal + vertical + resting components) is as follows:

$$\dot{V}O_2 = (0.1 \text{ mL·kg}^{-1}\text{·m}^{-1} \times \text{speed})$$
$$+ (1.8 \text{ mL·kg}^{-1}\text{·m}^{-1} \times \text{speed} \times \text{grade})$$
$$+ 3.5 \text{ mL·kg}^{-1}\text{·min}^{-1}$$

This equation will give the most accurate prediction of $\dot{V}O_2$ using speeds of 50 to 100 m·min^{-1} (1.9-3.7 mi·h^{-1}) while employing an incline during walking (Gibson, Wagner, and Heyward 2019). The first coefficient (0.1) represents the oxygen cost of moving our body mass per meter during horizontal walking. The second coefficient (1.8) represents the oxygen cost of moving our body mass per meter up an incline, or essentially against gravity. Finally, the units for speed are in m·min^{-1} (see the conversion factor for mi·h^{-1} to m·min^{-1} in table 6.1) and the grade is entered into the equation in fraction or decimal form. For example, if a person is walking at a 10% grade, 0.10 would be entered into the equations (10%/100 = 0.10).

Running

The ACSM running equation (horizontal + vertical + resting components) is as follows:

$$\dot{V}O_2 = (0.2 \text{ mL·kg}^{-1}\text{·m}^{-1} \times \text{speed})$$
$$+ (0.9 \text{ mL·kg}^{-1}\text{·m}^{-1} \times \text{speed} \times \text{grade})$$
$$+ 3.5 \text{ mL·kg}^{-1}\text{·min}^{-1}$$

TABLE 6.1 Common Unit Conversions

Length			
1 mi	1,609.34 m	1.609 km	5,280 ft
1 m	39.37 in.	3.281 ft	1.094 yd
1 ft	30.48 cm	0.3048 m	
1 km	0.621 mi		
1 in.	2.54 cm	0.0254 m	
Speed			
1 mi·h^{-1}	26.8 m·min^{-1}	1.609 km·h^{-1}	
1 km·h^{-1}	0.621 mi·h^{-1}		
Mass			
1 kg	2.205 lb		
1 lb	0.454 kg		
Energy, work, and power			
1 kgm	9.807 Nm or J	7.233 ft-lb	
1 kcal	4.184 kJ		
1 kJ	0.239 kcal		
1 MET	3.5 mL·kg^{-1}·min^{-1}		
1 W	6.12 kg·m·min^{-1}		
1 kg·m·min^{-1}	0.164 W		
1 L·min^{-1}	5 kcal·L^{-1} (estimated kcal burned for every L of oxygen consumed per min)		
Temperature			
Celsius to Fahrenheit	(°C × 1.8) + 32		
Fahrenheit to Celsius	(°F – 32) × 0.56		

This equation will give the most accurate prediction of $\dot{V}O_2$ using speeds greater than 134 m·min⁻¹ (>5 mi·h⁻¹). However, if the individual is truly "jogging" at a speed of <5 mi·h⁻¹, it is acceptable to apply this equation in cases where the speed is >3 mi·h⁻¹ (>80 m·min⁻¹). The first coefficient (0.2) represents the oxygen cost of moving our body mass per meter during horizontal running. Interestingly, when comparing the two horizontal coefficients of walking and running, running is twice (0.1 versus 0.2 mL·kg⁻¹·m⁻¹) the oxygen cost as walking. The second coefficient (0.9) represents the oxygen cost of moving our body mass per meter up an incline while running. Once again, the units for speed are in m·min⁻¹ and the grade is put into the equation in fraction or decimal form as was discussed with the walking equation.

Stepping

The ACSM stepping equation (horizontal + vertical + resting components) is as follows:

$$\dot{V}O_2 = (0.2 \text{ mL·kg}^{-1}\cdot\text{m}^{-1} \times \text{step rate}) + [1.33 \times (1.8 \text{ mL·kg}^{-1}\cdot\text{m}^{-1} \times \text{step height} \times \text{step rate})] + 3.5 \text{ mL·kg}^{-1}\cdot\text{min}^{-1}$$

This equation will give the most accurate prediction of $\dot{V}O_2$ using step rates of 12 to 30 steps·min⁻¹ and step heights of 1.6 to 15.7 in. (0.04-0.4 m). The units for step height are in meters. The first coefficient (0.2) represents the oxygen cost of moving our body mass per meter horizontally during stepping. The second coefficient (1.8) represents the oxygen cost of moving our body mass per meter up a step against gravity. The final coefficient (1.33) is a correction factor that accounts for the additional oxygen cost of stepping down (33% additional cost) as well as stepping up during this activity.

Leg Ergometry or Cycling

The ACSM leg ergometry or cycling equation (vertical + horizontal + resting components) is as follows:

$$\dot{V}O_2 = [(1.8 \text{ mL·kg}^{-1}\cdot\text{m}^{-1} \times \text{work rate})/\text{body mass}] + 3.5 \text{ mL·kg}^{-1}\cdot\text{min}^{-1} + 3.5 \text{ mL·kg}^{-1}\cdot\text{min}^{-1}$$

This equation will give the most accurate prediction of $\dot{V}O_2$ using work rates of 300 to 1,200 kg·m·min⁻¹ (50-200 W). The units for work rate and body mass are in kg·m·min⁻¹ and kg, respectively. The first coefficient (1.8) represents the oxygen cost of pedaling against resistance to produce 1

kilogram-meter of work. The next two components represent the oxygen cost of cycling against no resistance (3.5) and the resting component (3.5) as seen in previous equations.

An additional equation to be familiar with that can be used to calculate work rate (power output) on mechanically braked cycle ergometers (e.g., Monark cycle ergometer) is as follows (ACSM 2021; Gibson, Wagner, and Heyward 2019):

$$\text{work rate (kg·m·min}^{-1}) = \text{kg} \times \text{m·rev}^{-1} \times \text{rev·min}^{-1}$$

where kg is the resistance on the cycle ergometer flywheel, m·rev⁻¹ is the distance the flywheel travels (spins) per pedal revolution, and rev·min⁻¹ is the pedal frequency or revolutions per minute. If using a Monark cycle ergometer, the m·rev⁻¹ in this equation is 6 m·rev⁻¹, which is a constant value specific to this device.

Arm Ergometry or Cycling

The ACSM arm ergometry or cycling equation (vertical + resting components) is as follows:

$$\dot{V}O_2 = [(3.0 \text{ mL·kg}^{-1}\cdot\text{m}^{-1} \times \text{work rate})/\text{body mass}] + 3.5 \text{ mL·kg}^{-1}\cdot\text{min}^{-1}$$

This equation will give the most accurate prediction of $\dot{V}O_2$ using work rates of 150 to 750 kg·m·min⁻¹ (25-125 W). The units for work rate and body mass are in kg·m·min⁻¹ and kg, respectively. The first coefficient (3.0) represents the oxygen cost of arm cycling against resistance to produce 1 kilogram-meter of work. The only other component in this equation is the resting component, represented again by 3.5 mL·kg⁻¹·min⁻¹.

The work rate or power output equation can also be used with arm ergometry. If using a Monark arm ergometer, the m·rev⁻¹ in this equation would be 2.4 m·rev⁻¹, which is a constant value specific to this device.

Necessary Equipment or Materials

- Calculator
- Pencil and scratch paper for working through each problem
- Metabolic calculations problem sheet (set 1)
- Unit conversion table for review (table 6.1)
- Figures 6.1*a* and 6.1*b* for review

Procedures

Complete the metabolic calculations problem sheet (set 1) for this lab and give to a lab partner when you are done for final review prior to handing in this assignment. The answer key for this set of problems is at the end of the lab. A second set of problems (extra metabolic calculations problem sheet, set 2) is also included for you to practice on your own to improve your mastery of the metabolic calculations. It takes a good amount of practice to become competent with the steps and mechanics of these calculations, so the more familiarity you have, the better off you will be.

Sample Problems for Metabolic Calculations

Sample calculations using the ACSM metabolic equations follow. Please review these examples before starting your calculation sets for this lab and calculate along with them to better understand the steps involved. Not every type of calculation problem that you will see in the calculation sets is covered in these examples; thus, you will be required to do some critical thinking and apply these metabolic equations in a few different situations.

Sample Walking Equation Problem

Calculate the $\dot{V}O_2$ and METs for a person who walks at 3.3 $mi\cdot h^{-1}$ and 12% grade on a treadmill:

1. Convert the speed in $mi\cdot h^{-1}$ to $m\cdot min^{-1}$.

$$3.3 \text{ mi}\cdot h^{-1} \times 26.8 = 88.44 \text{ m}\cdot min^{-1}$$

2. Solve for the horizontal and vertical components (resting component is included).

$$\dot{V}O_2 = (0.1 \text{ mL}\cdot kg^{-1}\cdot m^{-1} \times 88.44 \text{ m}\cdot min^{-1})$$
$$+ (1.8 \text{ mL}\cdot kg^{-1}\cdot m^{-1} \times 88.44 \text{ m}\cdot min^{-1} \times 0.12)$$
$$+ 3.5 \text{ mL}\cdot kg^{-1}\cdot min^{-1}$$

$$\dot{V}O_2 = (8.84 \text{ mL}\cdot kg^{-1}\cdot min^{-1}) + (19.1030 \text{ mL}\cdot kg^{-1}\cdot min^{-1})$$
$$+ 3.5 \text{ mL}\cdot kg^{-1}\cdot min^{-1}$$

3. Solve for $\dot{V}O_2$ by adding all components together.

$$\dot{V}O_2 = (8.84 \text{ mL}\cdot kg^{-1}\cdot min^{-1}) + (19.1030 \text{ mL}\cdot kg^{-1}\cdot min^{-1})$$
$$+ 3.5 \text{ mL}\cdot kg^{-1}\cdot min^{-1}$$

$$\dot{V}O_2 = 31.44 \text{ mL}\cdot kg^{-1}\cdot min^{-1}$$

4. Solve for METs.

$$\text{METs} = 31.44 \text{ mL}\cdot kg^{-1}\cdot min^{-1} \div 3.5 \text{ mL}\cdot kg^{-1}\cdot min^{-1}$$

$$\text{METs} = 9 \text{ METs}$$

For this problem, remember to convert the grade into a decimal form (divide grade by 100) before putting it into the equation. Do not round your numbers until you get to the final answer for all of your calculations. $\dot{V}O_2$ can be rounded to the hundredth (two decimal places), while METs can be round to the tenth or up to a whole number. Also, be aware that you can solve for a speed or grade if given a $\dot{V}O_2$ value. Let's work through an example problem with that scenario.

Calculate the speed ($mi\cdot h^{-1}$) for a person who would like to walk at an intensity of 3.5 METs on a track (flat ground) as follows:

1. Convert METs into $\dot{V}O_2$.

$$\dot{V}O_2 = 3.5 \text{ METs} \times 3.5 \text{ mL}\cdot kg^{-1}\cdot min^{-1}$$

$$\dot{V}O_2 = 12.25 \text{ mL}\cdot kg^{-1}\cdot min^{-1}$$

2. Solve for speed. Subtract both sides by 3.5 $mL\cdot kg^{-1}\cdot min^{-1}$ first (order of operations).

$$12.25 \text{ mL}\cdot kg^{-1}\cdot min^{-1} = (0.1 \text{ mL}\cdot kg^{-1}\cdot m^{-1} \times \text{speed})$$
$$+ (1.8 \text{ mL}\cdot kg^{-1}\cdot m^{-1} \times \text{speed} \times 0) + 3.5 \text{ mL}\cdot kg^{-1}\cdot min^{-1}$$

$$8.75 \text{ mL}\cdot kg^{-1}\cdot min^{-1} = (0.1 \text{ mL}\cdot kg^{-1}\cdot m^{-1} \times \text{speed})$$
$$+ (1.8 \text{ mL}\cdot kg^{-1}\cdot m^{-1} \times \text{speed} \times 0)$$

3. Solve for speed (divide both sides by 0.1 $mL\cdot kg^{-1}\cdot m^{-1}$). The vertical component is zero.

$$8.75 \text{ mL}\cdot kg^{-1}\cdot min^{-1} = (0.1 \text{ mL}\cdot kg^{-1}\cdot m^{-1} \times \text{speed}) + (0)$$

$$87.5 \text{ m}\cdot min^{-1} = \text{speed}$$

4. Convert speed from $m\cdot min^{-1}$ to $mi\cdot h^{-1}$.

$$87.5 \text{ m}\cdot min^{-1} \div 26.8 = 3.3 \text{ mi}\cdot h^{-1}$$

Sample Running Equation Problem

Calculate the $\dot{V}O_2$ and METs for a person who runs at 8.5 $mi\cdot h^{-1}$ on a flat (zero grade) treadmill as follows:

1. Convert the speed in $mi\cdot h^{-1}$ to $m\cdot min^{-1}$.

$$8.5 \text{ mi}\cdot h^{-1} \times 26.8 = 227.8 \text{ m}\cdot min^{-1}$$

2. Solve for the horizontal and vertical components (resting component is included).

$$\dot{V}O_2 = (0.2 \text{ mL·kg}^{-1}\text{·m}^{-1} \times 227.8 \text{ m·min}^{-1})$$
$$+ (0.9 \text{ mL·kg}^{-1}\text{·m}^{-1} \times 227.8 \text{ m·min}^{-1} \times 0)$$
$$+ 3.5 \text{ mL·kg}^{-1}\text{·min}^{-1}$$

$$\dot{V}O_2 = (45.56 \text{ mL·kg}^{-1}\text{·min}^{-1}) + (0) + 3.5 \text{ mL·kg}^{-1}\text{·min}^{-1}$$

3. Solve for $\dot{V}O_2$ by adding all components together.

$$\dot{V}O_2 = (45.56 \text{ mL·kg}^{-1}\text{·min}^{-1}) + (0) + 3.5 \text{ mL·kg}^{-1}\text{·min}^{-1}$$

$$\dot{V}O_2 = 49.06 \text{ mL·kg}^{-1}\text{·min}^{-1}$$

4. Solve for METs.

$$\text{METs} = 49.06 \text{ mL·kg}^{-1}\text{·min}^{-1} \div 3.5 \text{ mL·kg}^{-1}\text{·min}^{-1}$$

$$\text{METs} = 14 \text{ METs}$$

Remember to convert the grade into a decimal form before putting it into the equation, just like with the walking equation. In this case, there is no vertical component due to running on flat ground; therefore, the vertical component is zero. And just like we saw with the walking equations, you can solve for a speed and grade if given a $\dot{V}O_2$ value. This can be further practiced in your problem sets.

Sample Stepping Equation Problem

Calculate the $\dot{V}O_2$ and METs for a person stepping at a rate of 22 steps·min^{-1} up and down a 30 cm (11.8 in.) step as follows:

1. Convert the step height from cm to m: 30 cm $\div$ 100 = 0.3 m.
2. Solve for the horizontal and vertical components (resting component is included).

$$\dot{V}O_2 = (0.2 \text{ mL·kg}^{-1}\text{·m}^{-1} \times 22 \text{ steps·min}^{-1})$$
$$+ [1.33 \, (1.8 \text{ mL·kg}^{-1}\text{·m}^{-1} \times 0.3 \text{ m} \times 22 \text{ steps·min}^{-1})]$$
$$+ 3.5 \text{ mL·kg}^{-1}\text{·min}^{-1}$$

$$\dot{V}O_2 = (4.4 \text{ mL·kg}^{-1}\text{·min}^{-1}) + (15.8004 \text{ mL·kg}^{-1}\text{·min}^{-1})$$
$$+ 3.5 \text{ mL·kg}^{-1}\text{·min}^{-1}$$

3. Solve for $\dot{V}O_2$ by adding all components together.

$$\dot{V}O_2 = (4.4 \text{ mL·kg}^{-1}\text{·min}^{-1}) + (15.8004 \text{ mL·kg}^{-1}\text{·min}^{-1})$$
$$+ 3.5 \text{ mL·kg}^{-1}\text{·min}^{-1}$$

$$\dot{V}O_2 = 23.7 \text{ mL·kg}^{-1}\text{·min}^{-1}$$

4. Solve for METs.

$$\text{METs} = 23.7 \text{ mL·kg}^{-1}\text{·min}^{-1} \div 3.5 \text{ mL·kg}^{-1}\text{·min}^{-1}$$

$$\text{METs} = 6.8 \text{ METs}$$

The most common error with this equation is failing to convert step height (inches or centime-ters) into meters; thus, do that first to make sure all numbers are correct at the start.

Sample Leg Cycling Equation Problem

Calculate the $\dot{V}O_2$, METs, and watt setting for a 75 kg (165 lb) person leg cycling at a work rate of 1,200 kg·m·min^{-1}:

1. Solve for the resistance component (horizontal and resting components are included). Multiply what is in parentheses first and then divide by body mass (order of operations).

$$\dot{V}O_2 = [(1.8 \text{ mL·kg}^{-1}\text{·m}^{-1} \times 1{,}200 \text{ kg·m·min}^{-1}) \div 75 \text{ kg}]$$
$$+ 3.5 \text{ mL·kg}^{-1}\text{·min}^{-1} + 3.5 \text{ mL·kg}^{-1}\text{·min}^{-1}$$

$$\dot{V}O_2 = (28.8 \text{ mL·kg}^{-1}\text{·min}^{-1}) + 3.5 \text{ mL·kg}^{-1}\text{·min}^{-1}$$
$$+ 3.5 \text{ mL·kg}^{-1}\text{·min}^{-1}$$

2. Solve for $\dot{V}O_2$ by adding all components together.

$$\dot{V}O_2 = (28.8 \text{ mL·kg}^{-1}\text{·min}^{-1}) + 3.5 \text{ mL·kg}^{-1}\text{·min}^{-1}$$
$$+ 3.5 \text{ mL·kg}^{-1}\text{·min}^{-1}$$

$$\dot{V}O_2 = 35.8 \text{ mL·kg}^{-1}\text{·min}^{-1}$$

3. Solve for METs.

$$\text{METs} = 35.8 \text{ mL·kg}^{-1}\text{·min}^{-1} \div 3.5 \text{ mL·kg}^{-1}\text{·min}^{-1}$$

$$\text{METs} = 10.2 \text{ METs}$$

4. Solve for watts.

$$\text{Watts} = 1{,}200 \text{ kg·m·min}^{-1} \div 6.12 \text{ kg·m·min}^{-1}$$

$$\text{Watts} = 196.1 \text{ W}$$

Watts and work rate in kg·m·min^{-1} can be rounded to the nearest tenth. Just like we saw with previous metabolic equations, if given a $\dot{V}O_2$ value, you can solve for work rate. Let's work through an example problem with that scenario.

Calculate a work rate for a cycle ergometer (in both kg·m·min^{-1} and watts) for an intensity of 8 METs (body mass is 75 kg [165 lb]) as follows:

1. Convert METs into $\dot{V}O_2$.

$$\dot{V}O_2 = 8 \text{ METs} \times 3.5 \text{ mL·kg}^{-1}\text{·min}^{-1}$$

$$\dot{V}O_2 = 28 \text{ mL·kg}^{-1}\text{·min}^{-1}$$

2. Solve for work rate. Subtract both sides by 7 mL·kg^{-1}·min^{-1} first (horizontal + resting component).

$$28 \text{ mL·kg}^{-1}\text{·min}^{-1} = [(1.8 \text{ mL·kg}^{-1}\text{·m}^{-1} \times \text{work rate})$$
$$\div 75 \text{ kg}] + 7 \text{ mL·kg}^{-1}\text{·min}^{-1}$$

$$21 \text{ mL·kg}^{-1}\text{·min}^{-1}$$
$$= [(1.8 \text{ mL·kg}^{-1}\text{·m}^{-1} \times \text{work rate}) \div 75 \text{ kg}]$$

3. Solve for work rate by multiplying both sides by kg body mass and then dividing by 1.8 mL·kg^{-1}·m^{-1}.

$$21 \text{ mL·kg}^{-1}\text{·min}^{-1}$$
$$= [(1.8 \text{ mL·kg}^{-1}\text{·m}^{-1} \times \text{work rate}) \div 75 \text{ kg}]$$

$$1{,}575 \text{ mL·min}^{-1} = (1.8 \text{ mL·kg}^{-1}\text{·m}^{-1} \times \text{work rate})$$

$$875 \text{ kg·m·min}^{-1} = \text{work rate}$$

4. Solve for watts.

$$\text{Watts} = 875 \text{ kg·m·min}^{-1} \div 6.12 \text{ kg·m·min}^{-1}$$

$$\text{Watts} = 143 \text{ W}$$

Sample Arm Cycling Equation Problem

Calculate the $\dot{V}O_2$, METs, and watt setting for a 75 kg (165 lb) person arm cycling at a work rate of 400 kg·m·min^{-1}:

1. Solve for the resistance component (resting component is included). Multiply what is in parentheses first and then divide by body mass.

$$\dot{V}O_2 = [(3.0 \text{ mL·kg}^{-1}\text{·m}^{-1} \times 400 \text{ kg·m·min}^{-1})$$
$$\div 75 \text{ kg}] + 3.5 \text{ mL·kg}^{-1}\text{·min}^{-1}$$

$$\dot{V}O_2 = (16 \text{ mL·kg}^{-1}\text{·min}^{-1}) + 3.5 \text{ mL·kg}^{-1}\text{·min}^{-1}$$

2. Solve for $\dot{V}O_2$ by adding all components together.

$$\dot{V}O_2 = (16 \text{ mL·kg}^{-1}\text{·min}^{-1}) + 3.5 \text{ mL·kg}^{-1}\text{·min}^{-1}$$

$$\dot{V}O_2 = 19.5 \text{ mL·kg}^{-1}\text{·min}^{-1}$$

3. Solve for METs.

$$\text{METs} = 19.5 \text{ mL·kg}^{-1}\text{·min}^{-1} \div 3.5 \text{ mL·kg}^{-1}\text{·min}^{-1}$$

$$\text{METs} = 5.8 \text{ METs}$$

4. Solve for watts.

$$\text{Watts} = 400 \text{ kg·m·min}^{-1} \div 6.12 \text{ kg·m·min}^{-1}$$

$$\text{Watts} = 65.4 \text{ W}$$

Remember that you can solve for work rate if given a $\dot{V}O_2$ value or for work rate (power output) if given a resistance load (kg) and pedaling rate (rev·min^{-1}). This can be practiced in your problem sets for both arm and leg cycling.

Sample Problems for Calculation of Energy Expenditure

Before starting your calculation sets, review the steps included in figures 6.1a and 6.1b. Be comfortable with calculating energy expenditure if given a MET level, $\dot{V}O_2$ value, or even a speed

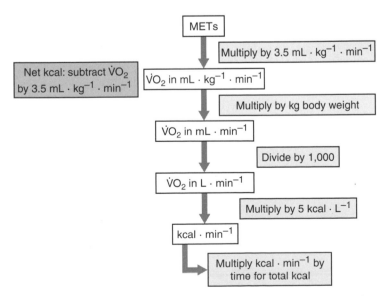

FIGURE 6.1a Energy expenditure calculation beginning with METs. Conversion factors are included in the boxes to the right of each variable.

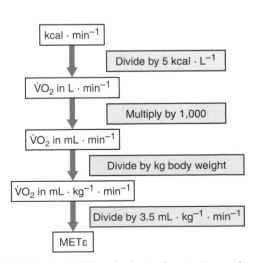

FIGURE 6.1b METs calculation beginning with energy expenditure. Conversion factors are included in the boxes to the right of each variable.

and grade on a treadmill or work rate on a cycle ergometer. Additionally, be prepared to calculate MET levels, $\dot{V}O_2$ values, or workloads if given an energy expenditure value. This is a very practical application of these metabolic calculation principles. For instance, if you choose a speed that you want your client to run at, you should be able to calculate an energy expenditure for them. Likewise, if a client asks you how fast they should walk if they want to burn 200 calories in an hour, you should be able to calculate that too. Let's practice a few of these scenarios.

Calculation sample 1:

A client (52 kg [114.4 lb]) is running (flat ground) at 6 METs. How many net kcal do they expend per minute and in 45 min?

1. Convert METs into $\dot{V}O_2$.

$$\dot{V}O_2 = 6 \text{ METs} \times 3.5 \text{ mL·kg}^{-1}\text{·min}^{-1}$$
$$\dot{V}O_2 = 21 \text{ mL·kg}^{-1}\text{·min}^{-1}$$

2. Subtract $\dot{V}O_2$ by 3.5 mL·kg^{-1}·min^{-1} (this is done to determine net kcal).

$$\dot{V}O_2 = 21 \text{ mL·kg}^{-1}\text{·min}^{-1} - 3.5 \text{ mL·kg}^{-1}\text{·min}^{-1}$$
$$\dot{V}O_2 = 17.5 \text{ mL·kg}^{-1}\text{·min}^{-1}$$

3. Multiply by kg body weight and divide by 1,000.

$$\dot{V}O_2 = 17.5 \text{ mL·kg}^{-1}\text{·min}^{-1} \times 52 \text{ kg}$$
$$\dot{V}O_2 = 910 \text{ mL·min}^{-1} \div 1,000$$
$$\dot{V}O_2 = 0.910 \text{ L·min}^{-1}$$

4. Multiply by 5 kcal·L^{-1} to get kcal·min^{-1}.

$$\text{kcal·min}^{-1} = 0.910 \text{ L·min}^{-1} \times 5 \text{ kcal·L}^{-1}$$
$$\text{kcal·min}^{-1} = 4.55 \text{ kcal·min}^{-1}$$

5. Multiply kcal·min^{-1} by exercise time to get total kcal.

$$\text{Total kcal} = 4.55 \text{ kcal·min}^{-1} \times 45 \text{ min}$$
$$\text{Total kcal} = 205 \text{ kcal}$$

The concept of net kcal relates to the calculation of energy expenditure that is only influenced by exercise and not resting metabolic rate (3.5 mL·kg^{-1}·min^{-1}) (ACSM 2021). Once exercise $\dot{V}O_2$ is calculated, it can be corrected by subtracting out the resting component. Accordingly, any calculation of energy expenditure following this corrective step will be specific to only that produced through exercise. It is recommended that the net kcal concept be used when determining the exercise needed to achieve weekly (kcal·wk^{-1}) and per day (kcal·day^{-1})

or per session goals. This will be discussed further in lab 7, concerning the design of aerobic exercise programs. If the $\dot{V}O_2$ correction for net kcal is not done, then the energy expenditure calculated would be "gross kcal" (exercise + resting $\dot{V}O_2$).

Calculation sample 2:

A client (90 kg [198 lb]) is running on a track (flat ground) at a 7 min and 30 s per mile pace. How many net kcal do they expend per minute and in 30 min?

1. Convert time in seconds to a decimal.

$$30 \text{ s} \div 60 \text{ s·min}^{-1} = 0.5 \text{ min}$$

2. Convert running pace to speed in mi·h^{-1}.

$$\text{Speed} = 60 \text{ min·h}^{-1} \div 7.5 \text{ min·mi}^{-1}$$
$$\text{Speed} = 8 \text{ mi·h}^{-1}$$

3. Convert the speed in mi·h^{-1} to m·min^{-1}.

$$8 \text{ mi·h}^{-1} \times 26.8 = 214.4 \text{ m·min}^{-1}$$

4. Solve for the horizontal and vertical components (resting component is included).

$$\dot{V}O_2 = (0.2 \text{ mL·kg}^{-1}\text{·m}^{-1} \times 214.4 \text{ m·min}^{-1})$$
$$+ (0.9 \text{ mL·kg}^{-1}\text{·m}^{-1} \times 214.4 \text{ m·min}^{-1} \times 0)$$
$$+ 3.5 \text{ mL·kg}^{-1}\text{·min}^{-1}$$

$$\dot{V}O_2 = (42.88 \text{ mL·kg}^{-1}\text{·min}^{-1}) + (0) + 3.5 \text{ mL·kg}^{-1}\text{·min}^{-1}$$

5. Solve for $\dot{V}O_2$ by adding all components together.

$$\dot{V}O_2 = (42.88 \text{ mL·kg}^{-1}\text{·min}^{-1}) + (0) + 3.5 \text{ mL·kg}^{-1}\text{·min}^{-1}$$
$$\dot{V}O_2 = 46.38 \text{ mL·kg}^{-1}\text{·min}^{-1}$$

6. Subtract $\dot{V}O_2$ by 3.5 mL·kg^{-1}·min^{-1}.

$$\dot{V}O_2 = 46.38 \text{ mL·kg}^{-1}\text{·min}^{-1} - 3.5 \text{ mL·kg}^{-1}\text{·min}^{-1}$$
$$\dot{V}O_2 = 42.88 \text{ mL·kg}^{-1}\text{·min}^{-1}$$

7. Multiply by kg body weight and divide by 1,000.

$$\dot{V}O_2 = 42.88 \text{ mL·kg}^{-1}\text{·min}^{-1} \times 90 \text{ kg}$$
$$\dot{V}O_2 = 3,859.2 \text{ mL·min}^{-1} \div 1,000$$
$$\dot{V}O_2 = 3.8592 \text{ L·min}^{-1}$$

8. Multiply by 5 kcal·L^{-1} to get kcal·min^{-1}.

$$\text{kcal·min}^{-1} = 3.8592 \text{ L·min}^{-1} \times 5 \text{ kcal·L}^{-1}$$
$$\text{kcal·min}^{-1} = 19.296 \text{ kcal·min}^{-1}$$

9. Multiply kcal·min^{-1} by exercise time to get total kcal.

$$\text{Total kcal} = 19.296 \text{ kcal·min}^{-1} \times 30 \text{ min}$$
$$\text{Total kcal} = 579 \text{ kcal}$$

Calculation sample 3:

A client (65 kg [136.7 lb]) wants to burn 300 kcal (net) in 30 min of exercise. How fast should they run on a flat treadmill (mi·h⁻¹)?

1. Calculate kcal·min⁻¹.

$$\text{kcal·min}^{-1} = 300 \text{ total kcal} \div 30 \text{ min}$$
$$\text{kcal·min}^{-1} = 10 \text{ kcal·min}^{-1}$$

2. Convert kcal·min⁻¹ to $\dot{V}O_2$ in L·min⁻¹.

$$\dot{V}O_2 = 10 \text{ kcal·min}^{-1} \div 5 \text{ kcal·L}^{-1}$$
$$\dot{V}O_2 = 2 \text{ L·min}^{-1}$$

3. Multiply by 1,000 and divide by kg body weight.

$$\dot{V}O_2 = 2 \text{ L·min}^{-1} \times 1,000$$
$$\dot{V}O_2 = 2,000 \text{ mL·min}^{-1} \div 65 \text{ kg}$$
$$\dot{V}O_2 = 30.77 \text{ mL·kg}^{-1}\text{·min}^{-1}$$

4. Solve for speed (divide both sides by 0.2 mL·kg⁻¹·m⁻¹). The vertical component is zero.

$$30.77 \text{ mL·kg}^{-1}\text{·min}^{-1} = (0.2 \text{ mL·kg}^{-1}\text{·m}^{-1} \times \text{speed})$$
$$+ (0.9 \text{ mL·kg}^{-1}\text{·m}^{-1} \times \text{speed} \times 0)$$
$$153.85 \text{ m·min}^{-1} = \text{speed}$$

5. Convert the speed in m·min⁻¹ to mi·h⁻¹.

$$153.85 \text{ m·min}^{-1} \div 26.8 = 5.7 \text{ mi·h}^{-1}$$

Note that in step 4, when solving for speed using the ACSM running equation, the resting component was omitted. This was because we are dealing with energy expenditure during exercise and these kcal are net kcal, not gross (includes resting $\dot{V}O_2$). In this case, speed is determined in the metabolic equation without the resting component included since this has already been subtracted out theoretically.

METABOLIC CALCULATIONS PROBLEM SHEET: SET 1

Complete this problem set and show all your work so you can better track your correct or incorrect solutions.

1. What is the $\dot{V}O_2$ of walking 3 mi·h^{-1} at 10% grade?

2. What is the $\dot{V}O_2$ of walking 2.5 mi·h^{-1} at 5% grade? Also, calculate the net kcal·min^{-1} for this workload and the net total kcal burned in 45 min (70 kg person).

3. What is the $\dot{V}O_2$ of running 6 mi·h^{-1} at 5% grade?

4. What is the $\dot{V}O_2$ of running 5 mi·h^{-1} at 2% grade? Also, calculate the net kcal·min^{-1} for this workload and the net total kcal burned in 50 min (70 kg person).

5. What is the $\dot{V}O_2$ of running 2.2 mi in 15 min? (Hint: Convert distance and time to mi·h^{-1} and solve.)

6. What is the MET level and net kcal expenditure for an 80 kg person who pedals 1,800 kg·m·min^{-1} on a cycle ergometer for 60 min?

7. What is the $\dot{V}O_2$ of stepping up and down an 8 in. step at a rate of 30 steps·min^{-1}?

8. Calculate running speed (mi·h^{-1} and mile pace in minutes and seconds) for a $\dot{V}O_2$ of 50 mL·kg^{-1}·min^{-1} (assuming flat ground).

9. Calculate a running speed (assuming flat ground) in mi·h^{-1} and mile pace (minutes and seconds) to meet a net kcal goal of 400 kcal for 40 min of exercise (60 kg person).

10. Calculate a walking speed (assuming flat ground) in mi·h^{-1} and mile pace (minutes and seconds) to meet a net kcal goal of 300 kcal for 60 min of exercise (90 kg person).

11. From a caloric expenditure standpoint, is it better to run a mile or walk a mile? Is it the same? Assume the client walks a mile at 3 mi·h^{-1} and runs at 6 mi·h^{-1} on flat ground and weighs 70 kg. Report your answer in *net total caloric expenditure* for the distance of a mile (not kcal·min^{-1}).

12. A client who utilizes a wheelchair (85 kg) is arm cycling at 4 METs. What is the watt setting for this individual and how many net total kcal would they expend in 25 min of exercise?

From J. Janot and N. Beltz, *Laboratory Assessment and Exercise Prescription* (Champaign, IL: Human Kinetics, 2023).

EXTRA METABOLIC CALCULATIONS PROBLEM SHEET: SET 2

Complete these problems on your own to further increase your comprehension and accuracy with unit conversions and metabolic calculations.

1. How many cm are there in 56 in.? _____

2. How many kg·m·min^{-1} is 125 W? _____

3. How many kg are there in 200 lb? _____

4. How many in. are there in 162 cm? _____

5. How many km are there in 5 mi? _____

6. How many mi are there in 35 km? _____

7. Walking at 3.1 mi·h^{-1} is equal to what m·min^{-1}? _____

8. Running at 6.0 mi·h^{-1} is equal to what m·min^{-1}? _____

9. Walking at 92 m·min^{-1} is equal to what mi·h^{-1}? _____

10. Running at 185 m·min^{-1} is equal to what mi·h^{-1}? _____

11. 8 METs is equivalent to what mL·kg^{-1}·min^{-1}? _____

12. How many METs are there in 32 mL·kg^{-1}·min^{-1}? _____

13. A person is leg cycling at a resistance of 2 kg on a Monark cycle ergometer at 50 rev·min^{-1}. What is the power output in kg·m·min^{-1}? _____

14. A person (75 kg) is arm cycling at a resistance of 1 kg on a Monark arm ergometer at 50 rev·min^{-1}. What is the power output in kg·m·min^{-1} and watts and what is the $\dot{V}O_2$ for the work rate? _____

15. An 81 kg person is exercising at 10 METs. What is their $\dot{V}O_2$ (mL·kg^{-1}·min^{-1}) and net kcal·min^{-1}? _____

16. Lance Lunge (75 kg) jogs 2 mi in 22 min and 30 s. How many mi·h^{-1} did Lance jog on average? _____

17. Jasmine Jumping-Jack (56 kg) is walking at 3.7 mi·h^{-1} at a 4% grade. What is Jasmine's net kcal expenditure in kcal·min^{-1}? _____

18. Darrell Dumbbell (66 kg) is running at 7 mi·h^{-1} on flat ground. What is Darrell's net kcal expenditure in kcal·min^{-1}? _____

19. Fernando Flexibility (150 lb) is running at 12 METs. What is Fernando's net kcal·min^{-1}? How fast is he running (flat ground)? _____

20. Cassie Curl-Up (120 lb) is cycling at 24 mL·kg^{-1}·min^{-1}. How many kcal (net) will Cassie expend in 25 min? _____

21. Bob Biceps (75 kg) is riding on a cycle ergometer at 120 W. What is the $\dot{V}O_2$ in METS, how many net kcal are burned in 30 min, and if Bob chooses to walk at the same intensity, how fast (in mi·h^{-1}) should Bob walk using a 10% grade? _____

22. Tony Triceps (85 kg) would like to expend 7 kcal·min^{-1} (gross kcal) while riding the cycle ergometer. What is the workload in kg·m·min^{-1}? _____

23. Connie Calorie (56 kg) wants to expend 10 kcal·min^{-1} (net kcal) during regular workouts. How fast should Connie run (flat ground)? _____

24. Lisa Lean (50 kg) is using a 12 in. step during her workout. If Lisa steps at a rate of 25 steps·min^{-1}, what is the $\dot{V}O_2$? What would the net kcal expenditure be in 40 min? _____

25. Peggy Pull-Up (75 kg) walks at 3.3 mi·h^{-1} on flat ground. How long does Peggy need to walk to burn 300 kcal (net)? _____

26. Gary Gains has a measured $\dot{V}O_2$max of 73 mL·kg^{-1}·min^{-1}. The measured ventilatory threshold is equivalent to 80% of the $\dot{V}O_2$max (58.4 mL·kg^{-1}·min^{-1}). How fast can Gary run (mi·h^{-1} and mile pace in minutes and seconds) at the ventilatory threshold (assuming flat ground)? _____

27. Nick Nautilus (70 kg) would like to know what running speed is needed (assuming flat ground) in mi·h^{-1} and mile pace (minutes and seconds) to meet a net kcal goal of 500 kcal for 50 min of exercise.

_____

28. Chantal Chin-Up (60 kg) would like to know what walking speed is needed (assuming flat ground) in mi·h^{-1} and mile pace (minutes and seconds) to meet a net kcal goal of 150 kcal for 60 min of exercise.

_____

From J. Janot and N. Beltz, *Laboratory Assessment and Exercise Prescription* (Champaign, IL: Human Kinetics, 2023).

METABOLIC CALCULATIONS PROBLEM SHEETS ANSWER KEYS

Answer Key for Metabolic Calculations Problem Sheet: Set 1

1. $26.01 \text{ mL·kg}^{-1}\text{·min}^{-1}$
2. $16.43 \text{ mL·kg}^{-1}\text{·min}^{-1}$; $4.53 \text{ kcal·min}^{-1}$; 204 total kcal
3. $42.9 \text{ mL·kg}^{-1}\text{·min}^{-1}$
4. $32.71 \text{ mL·kg}^{-1}\text{·min}^{-1}$; $10.22 \text{ kcal·min}^{-1}$; 511 total kcal
5. $50.67 \text{ mL·kg}^{-1}\text{·min}^{-1}$
6. 13.6 METs; 1,056 total kcal
7. $24.09 \text{ mL·kg}^{-1}\text{·min}^{-1}$
8. 8.7 mi·h^{-1}; ~6:53 min pace
9. 6.2 mi·h^{-1}; ~9:40 min pace
10. 4.1 mi·h^{-1}; ~14:38 min pace
11. ~56 total kcal for walking; ~113 kcal for running
12. ~49 W; ~112 total kcal

Answer Key for Extra Metabolic Calculations Problem Sheet: Set 2

1. 142.24 cm
2. $765 \text{ kg·m·min}^{-1}$
3. 90.7 kg
4. 63.8 in.
5. 8.1 km
6. 21.8 mi
7. 83.1 m·min^{-1}
8. 160.8 m·min^{-1}
9. 3.4 mi·h^{-1}
10. 6.9 mi·h^{-1}
11. $28 \text{ mL·kg}^{-1}\text{·min}^{-1}$
12. 9.1 METs
13. $600 \text{ kg·m·min}^{-1}$
14. $120 \text{ kg·m·min}^{-1}$; ~20 W; $8.3 \text{ mL·kg}^{-1}\text{·min}^{-1}$
15. $35 \text{ mL·kg}^{-1}\text{·min}^{-1}$; $2.56 \text{ kcal·min}^{-1}$
16. 5.3 mi·h^{-1}
17. $4.78 \text{ kcal·min}^{-1}$
18. $12.38 \text{ kcal·min}^{-1}$
19. $13.1 \text{ kcal·min}^{-1}$; 7.2 mi·h^{-1}
20. 140 total kcal
21. 7 METs; 238 total kcal; 2.82 mi·h^{-1}
22. $447.2 \text{ kg·m·min}^{-1}$
23. 6.7 mi·h^{-1}
24. $27.99 \text{ mL·kg}^{-1}\text{·min}^{-1}$; 245 total kcal
25. ~91 min
26. 10.2 mi·h^{-1}; ~5:53 min pace
27. 5.3 mi·h^{-1}; ~11:19 min pace
28. 3.1 mi·h^{-1}; ~19:21 min pace

From J. Janot and N. Beltz, *Laboratory Assessment and Exercise Prescription* (Champaign, IL: Human Kinetics, 2023).

Assessment of Aerobic Fitness: Submaximal Exercise Testing

Purpose of the Lab

In lab 6, we established that oxygen consumption ($\dot{V}O_2$) is the best measure of cardiorespiratory or aerobic fitness during exercise (Astrand et al. 2003). The evaluation of cardiorespiratory fitness in a comprehensive health and fitness assessment is done for several reasons. First, it is important to establish a baseline cardiorespiratory fitness level so exercise professionals can determine the most appropriate starting point for frequency, intensity, time, and type (FITT) prescriptions prior to beginning an aerobic exercise program. Second, cardiorespiratory fitness is considered an important risk factor for premature, all-cause mortality by organizations such as the American Heart Association and the American College of Sports Medicine (ACSM) plus several research groups (ACSM 2021; Barry et al. 2014; Blair et al. 1989; Kaminsky et al. 2019; Lee et al. 2010; Myers et al. 2015; Ross et al. 2016). An interventional strategy to reduce premature mortality risk, like starting an aerobic exercise program, can be recommended if

this information is known to exercise and medical professionals. Finally, cardiorespiratory fitness testing can be employed to assess fitness changes that occur over a period of exercise training. It is often advantageous for the exercise professional to gauge how their client is adapting over time to the stress of an aerobic fitness program and adjust the plan if needed. Also, it is very important to evaluate the effectiveness of the overall program by assessing whether cardiorespiratory fitness goals were reached at the conclusion of the program. All these reasons for performing cardiorespiratory fitness testing are consistent across a variety of populations including age, sex, fitness level, and disease state.

If we measure $\dot{V}O_2$ during incremental exercise up to maximal effort, it is termed $\dot{V}O_2$max. $\dot{V}O_2$max is defined as the maximal rate at which an individual can consume oxygen during exercise, usually detected as a plateau in $\dot{V}O_2$ with increasing workload (Robergs and Roberts 1997). It is important to note that the unit of oxygen consumption is described as a rate because the volume of oxygen consumed is always per unit of time (i.e., $L \cdot min^{-1}$, $mL \cdot min^{-1}$, and $mL \cdot kg^{-1} \cdot min^{-1}$). While $\dot{V}O_2max$ is an important determinant of fitness for our clients, measuring it through the best method possible (indirect calorimetry using open circuit spirometry as shown in lab 8) can be expensive and technically challenging in some situations. Thus, we need a more feasible, and often less strenuous, alternative to quantify this variable. This is where the use of submaximal exercise testing methods to measure $\dot{V}O_2max$ has its advantages.

The primary purpose of this lab is to introduce you to the methods involved in predicting $\dot{V}O_2max$ during submaximal exercise. In this lab, you will learn what factors might influence the decision to perform either a submaximal or a maximal cardiorespiratory exercise test and the assumptions behind the prediction of $\dot{V}O_2max$ from submaximal exercise tests. Plus, you will be provided with opportunities to perform and administer submaximal cardiorespiratory exercise tests and determine $\dot{V}O_2max$ from select prediction equations. During these experiences, you will also practice gathering heart rate (HR) and blood pressure (BP) at rest and during exercise plus monitoring your client's rating of perceived exertion (RPE) during exercise. The secondary purpose of this lab is to teach the principles behind the design of an aerobic exercise program using the FITT method recommended by ACSM (2021). A case study will be given to provide you with an opportunity to apply these principles along with how we might use the ACSM metabolic equations learned in lab 6 within the aerobic exercise prescription.

Choosing a Submaximal or a Maximal Cardiorespiratory Exercise Test

There are a variety of exercise protocols and modalities that an exercise professional can choose from to test cardiorespiratory fitness in their client, but there are two basic categories that all exercise tests fall into: either submaximal or maximal testing. The decision to perform a maximal or submaximal cardiorespiratory exercise test on a client depends on a few factors that need consideration (ACSM 2021). The following list includes some questions to reflect upon if you are involved with making this decision:

1. What is the reason for exercise testing and what type of information do you want from the test? For example, the main reason for testing a particular person could be to diagnose underlying coronary heart disease, check the effectiveness of a treatment intervention (e.g., heart surgery) that a patient has had, or evaluate how the medications that they are taking to manage their disease impact their exercise tolerance. In these cases, a maximal exercise test is the best choice. In contrast, a client who is apparently healthy and either wants a general gauge of their cardiorespiratory fitness or desires this information as a baseline for an aerobic exercise prescription might be more suited for a quicker and simpler submaximal exercise test. Clearly, specific and sensible answers to these two questions are critical for making the correct choice on what type of exercise test to administer.

2. What type of client are you testing and in what setting is the testing being performed? If your client has a history of disease or is at higher risk for either the presence of disease or experiencing an adverse event during exercise testing, doing a maximal test to evaluate their functional capacity and exercise tolerance may be more appropriate. Additionally, the setting that this client should be tested in aligns more with a hospital, laboratory, or clinic. On the other hand, if you are testing apparently healthy clients and the setting is a health and fitness facility, submaximal exercise testing would be much more appropriate.

3. What type of equipment and personnel are available in the setting? As discussed previously, maximal exercise testing to evaluate cardiorespiratory fitness usually involves specialized and expensive equipment. Because of this, most maximal exercise tests are conducted in a hospital, laboratory, or clinical setting. Also, personnel (e.g., exercise physiologists, nurses, physical therapists, etc.) who conduct exercise tests in these settings are usually very well trained in using this equipment and experienced with administering these tests with individuals from special populations. In settings

where there are less specialized equipment and nonmedical exercise professionals like in a health and fitness facility, submaximal exercise testing is more appropriate. All that is usually needed to gather cardiorespiratory fitness information is a stopwatch, HR monitor, an exercise modality, and an exercise professional to conduct the test.

Basic Assumptions of Submaximal Cardiorespiratory Fitness Testing

The main objective of submaximal cardiorespiratory fitness testing is to quantify the HR response to one or more submaximal exercise workloads and predict $\dot{V}O_2$max from these responses (ACSM 2021). However, several assumptions must be made when we engage in submaximal exercise testing. Accuracy in the prediction of $\dot{V}O_2$max is best achieved when the following assumptions are met (ACSM 2021):

1. *A steady-state HR is achieved for each work rate throughout the exercise test.* Since the main objective of submaximal exercise testing is to predict $\dot{V}O_2$max using the HR response during exercise, it is especially key to measure HR accurately during exercise. Thus, this assumption is relatively easy to meet if HR is carefully measured and the exercise professional follows the exercise protocol guidelines. We will see that some protocols define the range needed to achieve steady-state HR and others follow the general recommendation of two consecutive HR measurements within 5 to 6 beats·min⁻¹ (ACSM 2021). Whichever guideline is used, make sure that the client achieves steady state for each stage of the protocol, or prediction error will increase.

2. *HR increases linearly with work rate (and $\dot{V}O_2$) during exercise. This relationship exists for all individuals.* Past research shows that many individuals do not demonstrate this relationship during incremental exercise, especially during incremental exercise to volitional fatigue (Bunc et al. 1995; Conconi et al. 1982; Hofmann et al. 1994; Hofmann et al. 1997; Vella and Robergs 2005). This is due

to the occurrence of biological variability in responses to exercise among individuals. Since regression equations used to calculate $\dot{V}O_2$max assume a linear relationship between HR and work rate, this can lead to prediction error in some clients; thus, interpretation of overall results should be made cautiously.

3. *Predicted maximal HR for a given age is uniform and is minimally different from the actual measured maximal HR.* The actual maximal HR mentioned here is the HR that is reached at the point of maximal effort during an incremental exercise test. Since that HR value is often not known, we must predict the maximal HR using the client's age. Predicted maximal HR is used along with submaximal HRs collected during exercise testing to predict $\dot{V}O_2$max. If the predicted maximal HR is close to the actual maximal HR, the submaximal test regression equation will predict $\dot{V}O_2$max much more accurately.

The most utilized maximal HR prediction formula is 220 − age (Fox, Naughton, and Haskell 1971). It is the simplest equation to calculate age-predicted maximal HR, but it is associated with a prediction error in the range of ±7 to 12 beats·min⁻¹ of the actual maximal HR (Robergs and Landwehr 2002; Sarzynski et al. 2013). For example, if your client is 20 years old, the 220 − age formula would calculate for them an age-predicted maximal HR of 200. The actual maximal HR (if measured) could vary anywhere from 188 to 212 at the highest prediction range. This relatively large range of error carries over to our calculation of $\dot{V}O_2$max, which can affect the accuracy of this result. Also, the prescription of aerobic exercise intensity using HR relies on accurate prediction of maximal HR, something that we will see later in this lab. Therefore, ACSM (2021) recommends the use of alternative equations to predict maximal HR. For this lab, we will use the following equation for the prediction of maximal HR (Gellish et al. 2007): HRmax = 207 − (0.7 × age). Using this equation will better meet this assumption compared to 220 − age.

4. *$\dot{V}O_2$ at a given work rate is similar for all individuals.* This assumption was addressed in the previous lab with the discussion of the ACSM metabolic equations and calculation of $\dot{V}O_2$ during steady-state exercise. This assumption points to the issue of mechanical efficiency, which will affect the $\dot{V}O_2$ elicited from exercise work rates. In this scenario, if a client is using an exercise modality that is not familiar or they have a movement limitation caused by low skill proficiency, injury, disability, and so forth, they will be less efficient and likely elicit a much higher $\dot{V}O_2$ for a given workload compared to someone who is more efficient. This variability could ultimately contribute to prediction error as work rate increases during the test; thus, again, interpretation of overall results should be made cautiously.

5. *The individual being tested is not taking any medication that can alter HR during exercise.* It is common to think of medications that can blunt the HR response during exercise (i.e., beta-blockers), but we should also consider that some medications can increase HR (i.e., sympathomimetics like albuterol inhalers). In cases where your client is taking medications that can affect the HR response to exercise, it is important to find an alternative means to predict $\dot{V}O_2$max that does not involve the measurement of HR. As a reference for you, please consult the list of medications in appendix A and review their effects on the exercise response prior to any submaximal exercise testing.

6. *The effect of other factors that could affect HR at rest and during exercise is minimal.* These factors could include caffeine consumption (especially if that individual is particularly sensitive to it or is consuming large quantities), poor health status (ill or acute disease state), smoking, anxiety, dehydration, or being in an environment that could increase HR, specifically a hot, humid environment.

In conclusion, although submaximal exercise testing is not as precise as maximal exercise testing regarding the measurement of $\dot{V}O_2$max, it still provides a reasonable estimate of cardiorespiratory fitness with less technical skill, equipment needs, and risk of potential complications during testing. However, it is important that the exercise professional strives to control measurement error as much as possible to maximize the accuracy of results. Performing at least two different submaximal exercise tests on a client and evaluating these results for agreement can assist with determining $\dot{V}O_2$max as well. Over time, repeated submaximal exercise tests can serve as a sensitive marker to monitor aerobic training effects, independent of $\dot{V}O_2$max determination. For instance, a lower HR and blood pressure response for a given workload during a test would be representative of a positive training response and improvement of cardiorespiratory fitness.

Necessary Equipment or Materials

- Calculator
- HR monitor
- Stethoscope and sphygmomanometer
- Stopwatch
- Cycle ergometer and treadmill (plus access to a 200 or 400 m track)
- RPE scale (6-20 scale)
- One copy of the cardiorespiratory fitness data sheet for recording results
- One copy of the aerobic exercise programming progression chart

Calibration of Equipment

Some of the modalities used in this lab, such as motorized treadmills and electronically braked cycle ergometers, cannot be calibrated within the exercise physiology lab setting. In cases where these devices are not working properly, the manufacturer would have to service them in order to restore their accuracy. The exception is the mechanically braked cycle ergometer, such as a Monark ergometer, in which a belt or rope provides resistance in the form of friction on the flywheel. The resistance is measured in terms of kilograms and can be applied through the loosening or tightening of the belt around the flywheel. If a known weight is attached to the belt system, an individual can measure whether the amount of resistance that the cycle is reading is equal to the weight that is attached. If the resistance on the cycle does not equal the weight, the cycle ergometer should be calibrated according to manufacturer guidelines.

Procedures

The main setup for the lab requires students to work in small groups (2-4 students) to collect data for each submaximal exercise test. Please record your own individual data, not a group member's data, on the data collection sheet for later analysis in this lab. You will complete three submaximal exercise tests: (1) a cycle ergometer test of your choice, (2) a treadmill test of your choice, and (3) a field test of your choice. Please compare the results of each of your submaximal exercise tests and examine for agreement among the overall results. An acceptable criterion for agreement is to obtain two measurements (or more) within 3 to 5 $mL\cdot kg^{-1}\cdot min^{-1}$ if seeking to average results for the evaluation of cardiorespiratory fitness.

Following testing and data analysis, you will work on the sample case study and design a basic 12 wk aerobic exercise program. Use the aerobic exercise programming progression chart to fill in the information that you choose to prescribe for the client based on their goals and demographical information. You will perform a variety of calculations, so keep your calculator handy!

General Process for Conducting a Submaximal Cardiorespiratory Fitness Test

This section outlines generalized procedures to follow before, during, and after a submaximal exercise test (ACSM 2021). In some cases, specific instructions may be given to follow for a particular exercise test. In these cases, the specific requirements are what should be followed since many are related to the overall validity of the test.

- *HR and BP measurement at rest and during exercise.* HR and BP should be measured to ensure that the client is in a resting state prior to beginning the exercise test. HR measurements during exercise are needed to establish a steady-state HR as part of the evaluation of cardiorespiratory fitness. Thus, HR may need to be measured multiple times per stage to check for steady-state attainment. BP should be monitored, if able, near the end of each stage to determine if it is responding appropriately to exercise stress.

- *RPE.* An RPE value should be gathered during the last minute of each stage. RPE can be an important tool to gauge the overall stress and exertion a client is experiencing during exercise. In this lab, we will use the original category scale (6-20) developed by Borg (1982). It is important to standardize your instructions when teaching your client how to use the RPE scale. First, explain that the scale is designed to evaluate the overall level of physical exertion experienced during exercise. Second, make sure your client can view the scale during all explanations and indicate what the bottommost number means (6 = resting) and the topmost number means (20 = maximal exertion). The client will determine what the other numbers represent specific to them during exercise. And, finally, instruct the client to focus on total body exertion and not discomfort or tiredness in one specific area. Familiarity and practice with the RPE scale over time will allow clients to better gauge their exertion during exercise.

- *Reasons for stopping a submaximal exercise test.* The most common reasons for stopping a submaximal exercise test are the test has been completed, the client reaches their top HR allowed (either 70% HR reserve or 85% HRmax), or the client elects to stop. The risk of an exercise-induced complication occurring during submaximal exercise is low, especially in the apparently healthy adult population. Even so, that does not preclude the exercise professional from closely monitoring signs and symptoms during the test. A test should be terminated if the client experiences chest pain, severe fatigue, dizziness and nausea, or shortness of breath, among other symptoms. Additionally, if BP responds by either decreasing (hypotensive response) or excessively increasing (hypertensive response) during exercise, the test should be terminated. Two other general reasons for stopping an exercise test are that the client fails to follow instructions and the equipment you are using stops working properly. Overall, the best way to stay abreast of signs and symptoms is to be in regular communication with your client and monitor them closely throughout the exercise test.

- *Warm-up and cool-down.* Unless stated specifically in the protocol procedures, the client should be given a 3 to 5 min warm-up on the modality. This will provide an opportunity for the client to prepare for the stress of exercise and become familiar with the modality (see the next bullet point) prior to starting the test. For the cool-down period, the work rate should be equal to or lower than the first stage work rate if the client is performing an active recovery, and should last for approximately

5 min. A passive recovery, with the client resting, is recommended if they are experiencing acute signs and symptoms, and the cool-down period should be extended for as long as required.

- *Other issues.* Clients should be given time to become familiar with the modality that they are using during the exercise test. Treadmills can be intimidating and tricky for some individuals who are not used to walking on them. Review with the client that, if they are having an issue like dizziness or losing balance, they can grab the handrails to steady themselves temporarily and stop if needed. For the bike, it is important to review with the client proper body posture (upright position and not slouched over), seat height positioning (slight bend at the knee at full extension of the leg and ball of foot on pedal), hand position and light hold on the handlebars, and pedaling at the required rev·min^{-1} or metronome beat prior to beginning the test.

Submaximal Cardiorespiratory Fitness Tests: Cycle Ergometer Tests

The two tests that will be discussed in this section are the modified YMCA cycle ergometer and the George submaximal cycle ergometer tests. Both of these tests provide a reasonable estimate of cardiorespiratory fitness and have been cross-validated to evaluate accuracy in multiple populations (Beekley et al. 2004; George et al. 2000).

Modified YMCA Cycle Ergometer Test

The modified YMCA cycle ergometer test is a multistage protocol designed to determine $\dot{V}O_2$max in both men and women (Golding 2000). The main goal of the test is to elicit a HR response between 110 beats·min^{-1} and 85% HRmax for at least two consecutive stages. Each stage is 3 min in length and the test has a maximum of four stages to complete to achieve the main HR response goal. To control for work rate, the pedal rate should be held constant at 50 rev·min^{-1} throughout the test. If there is no reading for rev·min^{-1} on the bike, set a metronome to 100 beats·min^{-1} and instruct your client to perform one pedal stroke per beat.

If you are using a mechanically braked bike like a Monark, set the first stage resistance to 0.5 kg to produce a work rate of 150 kg·m·min^{-1}. For each stage, measure the HR during the second and third minute (see the modified YMCA data collection table on the cardiorespiratory fitness data sheet for exact timing) and determine if the client has reached a steady state. If the HRs differ by more than 5 beats·min^{-1}, have the client complete another minute at that workload until HR agreement is achieved. RPE using the category scale (6-20) and BP should be collected at the time periods indicated in the modified YMCA data collection table.

The progression of work rate during the modified YMCA test is dictated by the HR response during the first stage of the test. The work rate in stage 2 should be set according to the following options (ACSM 2021):

Option 1: HR is >100 beats·min^{-1} → increase work rate to 300 kg·m·min^{-1} (1.0 kg resistance; 50 W)

Option 2: HR is 90 to 100 beats·min^{-1} → increase work rate to 450 kg·m·min^{-1} (1.5 kg resistance; 75 W)

Option 3: HR is 80 to 89 beats·min^{-1} → increase work rate to 600 kg·m·min^{-1} (2.0 kg resistance; 100 W)

Option 4: HR is <80 beats·min^{-1} → increase work rate to 750 kg·m·min^{-1} (2.5 kg resistance; 125 W)

For stage 3, increase the work rate by 150 kg·m·min^{-1} (0.5 kg; 25 W) after obtaining a steady-state HR in stage 2. If stage 4 is needed to achieve the goal of two consecutive HRs between 110 beats·min^{-1} and 85% HRmax, then increase the work rate by 150 kg·m·min^{-1} (0.5 kg; 25 W) again.

The calculation of $\dot{V}O_2$max is done using the multistage model, which includes $\dot{V}O_2$ and HR data from the last two stages completed in the modified YMCA test (Gibson, Wagner, and Heyward 2019). The first step is to calculate $\dot{V}O_2$ from the last two work rates completed in the test using the ACSM cycle ergometer metabolic equation from lab 6. Once you have these values, the next step is to calculate the slope (b) value using the $\dot{V}O_2$ and HR data gathered during the test:

$$b = (\dot{V}O_2SM_2 - \dot{V}O_2SM_1) / (HR_2 - HR_1)$$

where $\dot{V}O_2SM_2$ is the last stage $\dot{V}O_2$, $\dot{V}O_2SM_1$ is the second to last stage $\dot{V}O_2$, HR_2 is the last stage HR, and HR_1 is the second to last stage HR. The final step is to calculate $\dot{V}O_2max$ using the following equation:

$$\dot{V}O_2max = \dot{V}O_2SM_2 + [b\,(HRmax - HR_2)]$$

where HRmax is the age-predicted maximum HR from the Gellish et al. (2007) equation. The $\dot{V}O_2max$ should be reported in $mL\cdot kg^{-1}\cdot min^{-1}$ for this lab.

Here is a sample $\dot{V}O_2max$ calculation for the modified YMCA ergometer test. Calculate the $\dot{V}O_2max$ for a 40-year-old person (70 kg [154 lb]) using the following data:

Stage 2 work rate and HR: 300 $kg\cdot m\cdot min^{-1}$ and 125 $beats\cdot min^{-1}$

Stage 3 work rate and HR: 450 $kg\cdot m\cdot min^{-1}$ and 145 $beats\cdot min^{-1}$

1. Solve for age-predicted HRmax.

$$HRmax = 207 - (0.7 \times 40)$$

$$HRmax = 179\ beats\cdot min^{-1}$$

2. Solve for the $\dot{V}O_2$ in stage 2 ($\dot{V}O_2SM_1$).

$$\dot{V}O_2SM_1 = [(1.8\ mL\cdot kg^{-1}\cdot m^{-1} \times 300\ kg\cdot m\cdot min^{-1}) \div 70\ kg] + 3.5\ mL\cdot kg^{-1}\cdot min^{-1} + 3.5\ mL\cdot kg^{-1}\cdot min^{-1}$$

$$\dot{V}O_2SM_1 = (7.7\ mL\cdot kg^{-1}\cdot min^{-1}) + 3.5\ mL\cdot kg^{-1}\cdot min^{-1} + 3.5\ mL\cdot kg^{-1}\cdot min^{-1}$$

$$\dot{V}O_2SM_1 = 14.7\ mL\cdot kg^{-1}\cdot min^{-1}$$

3. Solve for the $\dot{V}O_2$ in stage 3 ($\dot{V}O_2SM_2$).

$$\dot{V}O_2SM_2 = [(1.8\ mL\cdot kg^{-1}\cdot m^{-1} \times 450\ kg\cdot m\cdot min^{-1}) \div 70\ kg] + 3.5\ mL\cdot kg^{-1}\cdot min^{-1} + 3.5\ mL\cdot kg^{-1}\cdot min^{-1}$$

$$\dot{V}O_2SM_2 = (11.6\ mL\cdot kg^{-1}\cdot min^{-1}) + 3.5\ mL\cdot kg^{-1}\cdot min^{-1} + 3.5\ mL\cdot kg^{-1}\cdot min^{-1}$$

$$\dot{V}O_2SM_2 = 18.6\ mL\cdot kg^{-1}\cdot min^{-1}$$

4. Solve for the slope.

$$b = (18.6\ mL\cdot kg^{-1}\cdot min^{-1} - 14.7\ mL\cdot kg^{-1}\cdot min^{-1}) / (145\ beats\cdot min^{-1} - 125\ beats\cdot min^{-1})$$

$$b = (3.9) / (20)$$

$$b = 0.195$$

5. Solve for $\dot{V}O_2max$.

$$\dot{V}O_2max = 18.6\ mL\cdot kg^{-1}\cdot min^{-1} + [0.195\,(179\ beats\cdot min^{-1} - 145\ beats\cdot min^{-1})]$$

$$\dot{V}O_2max = 18.6\ mL\cdot kg^{-1}\cdot min^{-1} + 6.63$$

$$\dot{V}O_2max = 25.2\ mL\cdot kg^{-1}\cdot min^{-1}$$

George Submaximal Cycle Ergometer Test

The George submaximal cycle ergometer test is a multistage protocol designed to determine $\dot{V}O_2max$ in apparently healthy men and women aged 18 to 39 years old (George et al. 2000). The main goal of the test is to elicit a HR response between 120 and 175 $beats\cdot min^{-1}$. Following a 2 min warm-up, 3 min stages of cycling exercise are completed to achieve the main HR response goal. To control for work rate, the pedal rate should be held constant at 70 $rev\cdot min^{-1}$ throughout the test. If there is no reading for $rev\cdot min^{-1}$ on the bike, set a metronome to 140 $beats\cdot min^{-1}$ and instruct your client to provide a pedal stroke on each beat.

If you are using a mechanically braked bike, set the warm-up resistance to 1.0 kg to produce a work rate of 420 $kg\cdot m\cdot min^{-1}$ (~68 W). Measure the HR response at the end of the warm-up to determine the workload for stage 1. For stage 1 (and 2 if needed), measure the HR during the second and third minute (see the George submaximal cycle ergometer test data collection table on the cardiorespiratory fitness data sheet for exact timing) and determine if the client has reached steady state. HRs should not differ by more than 6 $beats\cdot min^{-1}$ to achieve steady state. Continue this process until a steady-state HR between the goal range of 120 and 175 $beats\cdot min^{-1}$ is elicited. RPE using the category scale should be collected at the time periods indicated on the George submaximal cycle ergometer data collection table. BP can also be measured once per stage to monitor for an appropriate response to exercise.

Table 7.1 details the protocol to follow for the George submaximal cycle ergometer test (George et al. 2000). Choose a column based on the measured HR response following the 2 min warm-up. If HR is <120 $beats\cdot min^{-1}$, set the work rate according to your client's body weight as indicated in table 7.1 for stage 1. If HR falls within the range of 120 to 175 $beats\cdot min^{-1}$, continue with the work rate that was performed in the warm-up for stage 1. In either case, establish a steady-state HR and record. If the steady-state HR at the end of stage 1 is still <120 $beats\cdot min^{-1}$, continue on to stage 2 and set the work rate as indicated in table 7.1. Once a steady-state HR is obtained within the range of 120 to 175 $beats\cdot min^{-1}$, transition your client into the cool-down phase of the protocol.

To calculate $\dot{V}O_2max$, enter body weight in kg, final work rate in watts, final steady-state HR, and

TABLE 7.1   George Submaximal Cycle Ergometer Test Protocol

Warm-Up (2 min): Set work rate at 420 kg·m·min⁻¹ (1.0 kg resistance; ~68 W); collect HR at 2 min.	
If HR <120 beats·min⁻¹ at end of warm-up	If HR = 120-175 beats·min⁻¹ at end of warm-up
Stage 1 (3 min): 840 kg·m·min⁻¹ (2.0 kg resistance; ~137 W)ᵃ; OR 630 kg·m·min⁻¹ (1.5 kg resistance; ~103 W)ᵇ	**Stage 1 (3 min):** 420 kg·m·min⁻¹ (1.0 kg resistance; ~68 W)
Stage 2 (3 min): If HR is still <120 beats·min⁻¹ at end of stage 1, increase resistance by 0.5 kg.	
If HR is at steady state between 120 and 175 beats·min⁻¹ at the end of stage 1 or 2, proceed to a 3-5 min cooldown at ≤150 kg·m·min⁻¹ or 25 W.	

Note: Pedal rate should be held constant at 70 rev·min⁻¹.

ᵃWork rate for men with body weight ≥73 kg (~160 lb).

ᵇWork rate for all women and for men with body weight <73 kg.

Adapted from George et al. (2000).

age in the equation that follows. For sex, 1 will be entered for men and 0 will be entered for women:

$$\dot{V}O_2max = 85.447 + 9.104(sex) - 0.2676(age) - 0.4150(kg) + 0.1317(watts) - 0.1615(HR)$$

Submaximal Cardiorespiratory Fitness Tests: Treadmill Tests

The two tests that will be discussed in this section are the Ebbeling walking and the George jogging treadmill tests. Both of these tests provide a reasonable estimate of cardiorespiratory fitness and have been cross-validated to evaluate accuracy in multiple populations (Ebbeling et al. 1991; George et al. 1993b).

Ebbeling Walk Test

The Ebbeling walking test is a single-stage treadmill protocol designed to determine $\dot{V}O_2max$ in apparently healthy men and women aged 20 to 59 years old (Ebbeling et al. 1991). The main goal of the test is to elicit a steady-state HR response using an individualized walking pace between 2.0 and 4.5 mi·h⁻¹ (53.6-120.6 m·min⁻¹). The walking pace will be established during a 4 min warm-up period at 0% grade, and the client should be instructed to select a brisk pace. Once the warm-up is completed, increase the treadmill grade to 5% and have the client walk an additional 4 min. Measure HR near the end of the fourth minute of the walking

stage at 5% grade. Record two HRs during this time to establish steady-state agreement (within 5-6 beats·min⁻¹). You will enter speed in mi·h⁻¹, HR, and age in the equation that follows. For sex, 1 will be entered for men and 0 will be entered for women:

$$\dot{V}O_2max = 15.1 + 21.8(speed) - 0.327(HR) - 0.263(speed \times age) + 0.00504(HR \times age) + 5.48(sex)$$

George Jog Test

The George jogging test is a single-stage treadmill protocol designed to determine $\dot{V}O_2max$ in apparently healthy men and women aged 18 to 28 years old (George et al. 1993b). The main goal of the test is to elicit a steady-state HR response using an individualized jogging pace between 4.3 and 7.5 mi·h⁻¹ (115.2-201 m·min⁻¹). The jogging pace established should be comfortable and held constant during the 3 min test. Women should not exceed 6.5 mi·h⁻¹ (174.2 m·min⁻¹) and men should not exceed 7.5 mi·h⁻¹ (201 m·min⁻¹). Measure HR following 3 min of jogging and record two consecutive HRs (30 s apart) until steady-state agreement (≤3 beats·min⁻¹) is established. HR should not exceed 180 beats·min⁻¹ at any point during the test. You will enter speed in mi·h⁻¹, HR, and body weight in kg in the equation that follows. For sex, 1 will be entered for men and 0 will be entered for women:

$$\dot{V}O_2max = 54.07 - 0.1938(kg) + 4.47(speed) - 0.1453(HR) + 7.062(sex)$$

Submaximal Cardiorespiratory Fitness Tests: Field Tests

There are some advantages to performing a field test over a laboratory test. Field tests require little equipment to administer, they involve less of a time commitment for clients to perform, and they can be used to test large groups of individuals at the same time compared to laboratory tests. However, some disadvantages to performing a field test are that clients are less closely monitored (i.e., less frequent HR and no exercise BP measurement, no monitoring of signs and symptoms during exercise, etc.) and clients are often required to complete the test as quickly as possible, leading to the likelihood of vigorous exercise intensities. It may not be appropriate for some clients to engage in vigorous exercise, especially if not closely monitored, thus the health and fitness of clients should be screened appropriately prior to testing and an alternative sought if not deemed safe. Another disadvantage is that the predictive accuracy of field tests for $\dot{V}O_2$max can be highly variable from client to client compared to laboratory tests (Gibson, Wagner, and Heyward 2019).

The three tests that will be discussed in this section are the Rockport 1-mile (1.6 km) walk, the George 1-mile (1.6 km) run, and the Larsen 1.5-mile (2.4 km) run/walk field tests. These three tests provide a reasonable estimate of cardiorespiratory fitness and have been cross-validated to evaluate accuracy in multiple populations (George et al. 1993a; Kline et al. 1987; Taylor et al. 2002).

Rockport 1-Mile Walk Test

The Rockport 1-mile walk is a walking test designed to determine $\dot{V}O_2$max in apparently healthy men and women aged 20 to 69 years old (Kline et al. 1987). The main goal of the test is to walk 1 mi as fast as possible and then measure HR immediately following exercise and record. HR should be measured for 15 s, starting when the client stops walking. It is recommended to use a track (200 or 400 m) to set the 1 mi walking course. If an alternative course is used, make sure that it is flat ground. Allow your client a general 5 to 10 min warm-up prior to the test.

You will enter body weight in kg, HR, time, and age in the equation that follows. For time, convert the minutes and seconds into minutes only by

dividing seconds by 60. For example, a 15 min and 30 s total time would be converted to 15.5 min. For sex, 1 will be entered for men and 0 will be entered for women:

$$\dot{V}O_2max = 132.853 - 0.0769(kg) - 0.3877(age) + 6.315(sex) - 3.2649(time) - 0.1565(HR)$$

George 1-Mile Run Test

The George 1-mile run test is a running test designed to determine $\dot{V}O_2$max in apparently healthy men and women aged 18 to 29 years old (George et al. 1993a). The main goal of the test is to run 1 mi at a moderate or submaximal steady-state intensity level and complete the test in ≥8 min for men and ≥9 min for women at a HR ≤180 beats·min^{-1}. HR is measured immediately following exercise using a 15 s count and recorded on the data sheet. It is recommended to use a track (200 or 400 m) to set the 1 mi running course. If an alternative course is used, make sure that it is flat ground. Also, record the RPE level at the completion of the test and allow your client a general 2 to 3 min warm-up prior to the test.

You will enter body weight in kg, HR, and time in min in the equation that follows. For time, convert the minutes and seconds into minutes only by dividing seconds by 60. For example, a 12 min and 30 s total time would be converted to 12.5 min. For sex, 1 will be entered for men and 0 will be entered for women:

$$\dot{V}O_2max = 100.5 - 0.1636(kg) - 1.438(min) - 0.1928(HR) + 8.344(sex)$$

Larsen 1.5-Mile Run/Walk Test

The Larsen 1.5-mile run/walk test is a test designed to determine $\dot{V}O_2$max in apparently healthy men and women aged 18 to 29 years old (Larsen et al. 2002). The main goal of the test is to walk or run 1.5 mi as fast as possible while maintaining a consistent pace equivalent to an RPE of 13 on the category scale ("somewhat hard"). Additionally, clients should maintain their HR between 60% and 90% HRmax during the test and use a HR monitor to assist with staying within this range. HR and RPE level should be measured immediately following exercise and recorded on the data sheet. It is recommended to use a track (200 or 400 m) to set the 1.5 mi running course. If an alternative course is used, make sure that it is flat ground.

Allow your client a general 2 to 3 min warm-up prior to the test.

You will enter body weight in kg, HR, and time in minutes in the equation that follows. For time, convert the minutes and seconds into minutes only by dividing seconds by 60. For example, a 10 min and 30 s total time would be converted to 10.5 min. For sex, 1 will be entered for men and 0 will be entered for women:

$$\dot{V}O_2max = 100.162 + 7.301(sex) - 0.164(kg) - 1.273(min) - 0.156(HR)$$

Normative Data and Interpretation

Tables 7.2 and 7.3 contain the normative $\dot{V}O_2max$ data values that are specific to men and women of varying age groups. You will use these normative charts for this lab and lab 8 to compare and classify your $\dot{V}O_2max$ data gathered in both labs. Table 7.2 is used for both the treadmill-based protocols and the running or walking field tests, and table 7.3 is used for the cycle ergometer–based protocols completed in this lab.

To use these tables, find your sex-specific section and age-specific row and then determine where your $\dot{V}O_2max$ value falls according to the ratings at the top of the chart. For this lab, classify your

$\dot{V}O_2max$ for all tests completed and compare your results. Typically, if there is good agreement in $\dot{V}O_2max$ values between two different submaximal exercise tests, these values can be averaged to get a representative $\dot{V}O_2max$ value for the client.

While it is important to classify and compare your client's $\dot{V}O_2max$ to those of similar age and sex, it is also very important to interpret this value with a health risk appraisal focus in mind. As discussed earlier in this lab, maintaining both cardiorespiratory fitness and physical activity across the lifespan is associated with a lower risk of morbidity and premature mortality. Many chronic health issues such as cardiovascular disease, metabolic diseases like type 2 diabetes, osteoporosis, and certain types of cancer are linked to poor physical activity and fitness (Bull et al. 2020; CDC 2021; Garber et al. 2011). Older adults also benefit from higher levels of physical activity and cardiorespiratory fitness by being able to maintain independence, quality of life, and overall physical and cognitive function. Thus, the overall discussion of this fitness information with your client should include these talking points: (1) where they are currently at from a fitness standpoint compared to others (classification), (2) what these data mean for their overall health risk appraisal (interpretation), (3) whether they should either maintain or improve their current fitness level (interpretation), and (4)

TABLE 7.2   Cardiorespiratory Fitness ($\dot{V}O_2max$) Ratings for Adults by Age and Sex: Treadmill

Age (yr)	Poor	Fair	Good	Excellent	Superior
MEN					
20-29	≤44.9	45-49.9	50-56.9	57-62.9	≥63
30-39	≤39.9	40-44.9	45-49.9	50-57.9	≥58
40-49	≤35.9	36-39.9	40-45.9	46-52.9	≥53
50-59	≤30.9	31-34.9	35-40.9	41-46.9	≥47
60-69	≤26.9	27-29.9	30-35.9	36-40.9	≥41
WOMEN					
20-29	≤33.9	34-39.9	40-45.9	46-51.9	≥52
30-39	≤27.9	28-31.9	32-36.9	37-40.9	≥41
40-49	≤24.9	25-28.9	29-32.9	33-38.9	≥39
50-59	≤21.9	22-24.9	25-28.9	29-32.9	≥33
60-69	≤18.9	19-21.9	22-24.9	25-27.9	≥28

Note: All values are in mL·kg⁻¹·min⁻¹ and were gathered using treadmill-based cardiopulmonary exercise testing (CPXT).

Adapted from Kaminsky, Arena, and Myers (2015).

TABLE 7.3   Cardiorespiratory Fitness ($\dot{V}O_2$max) Ratings for Adults by Age and Sex: Cycle Ergometer

Age (yr)	Poor	Fair	Good	Excellent	Superior
MEN					
20-29	≤37.9	38-43.9	44-50.9	51-57.9	≥58
30-39	≤27.9	28-30.9	31-35.9	36-43.9	≥44
40-49	≤24.9	25-28.9	29-33.9	34-40.9	≥41
50-59	≤23.5	23.6-25.9	26-29.9	30-36.9	≥37
60-69	≤20.9	21-22.9	23-25.9	26-30.9	≥31
WOMEN					
20-29	≤27.9	28-32.9	33-37.9	38-44.9	≥45
30-39	≤19.9	20-22.9	23-25.9	26-32.9	≥33
40-49	≤17.9	18-20.9	21-22.9	23-28.9	≥29
50-59	≤16.4	16.5-17.9	18.0-20.9	21-24.9	≥25
60-69	≤15.2	15.3-16.6	16.7-18.9	19-21.9	≥22

Note: All values are in mL·kg^{-1}·min^{-1} and were gathered using cycle ergometer–based CPXT.

Adapted from Kaminsky et al. (2017).

what they should do to achieve their goals (interpretation and planning of exercise program). By focusing on these points, you will be able to make the biggest impact on your client's understanding of the data and help them plan an appropriate exercise program.

Exercise Program Design: Aerobic Exercise Prescription

The main purpose of the aerobic exercise prescription is to provide a gradual exercise stimulus over time to improve the cardiorespiratory fitness of your client. Other individual goals in an aerobic exercise program could include decreasing the risk of cardiovascular and other chronic diseases, overall weight loss, trying new modes of exercise, or providing an opportunity for social interaction if exercising with other individuals. Whatever the goals are, it is the responsibility of the exercise professional to encourage clients to set their own goals, assist clients in further clarifying and defining those goals, and then design an individualized exercise program with these goals in mind. Remember, individualization of any exercise program design is key to promote adherence to the program and a personal behavior change in your client. This section will provide you with information to design a basic aerobic exercise program for clients seeking to improve their health and fitness.

Overall, the aerobic exercise program should include four main components: (1) warm-up, (2) aerobic exercise training, (3) recreational activity (optional), and (4) cool-down. The warm-up period should be designed to get the body ready for aerobic activity and lasts approximately 5 to 10 min. The modality chosen for the warm-up can be anything, but it is usually specific to the exercise modality chosen for the exercise training session. The cool-down period is needed to allow for HR, BP, body temperature, breathing rate, and so forth, to gradually return to resting levels. This period involves movement either specific to the modality used during exercise or involves a different exercise altogether. Typically, a cool-down is performed for 5 to 10 min as well. The recreational component includes activities that can be used to supplement the aerobic training phase. Activities such as basketball, rollerblading, soccer, outdoor Nordic skiing, or racket sports like tennis can provide an additional aerobic training stimulus on days when traditional endurance activities (e.g., walking, running, cycling, etc.) are not being performed. This is also an effective way to provide cross-training opportunities for clients to engage in to reduce the risk of overtraining and overuse injuries. It is recommended that clients should be of average fitness for their age and sex and participate in a regular, aerobic exercise program prior to engaging in recreational activities (Gibson, Wagner, and Heyward 2019). The final component, aerobic exercise program design, is covered in the following sections.

Frequency

Frequency of exercise refers to the number of days a client exercises per week. ACSM (2021) recommends an exercise training frequency of at least 3 $d \cdot wk^{-1}$ and up to 5 $d \cdot wk^{-1}$ depending on factors such as goals, time commitment, fitness level, and target training intensity, to name a few.

Intensity

It could be argued that intensity is the most important prescription variable influencing both cardiorespiratory fitness improvements and overall adherence to an exercise program (ACSM 2021). For example, if the exercise professional is too aggressive and overprescribes exercise intensity, the client may become overtrained and experience overuse injuries that could affect both their ability to and enjoyment of exercise. Alternatively, if the exercise professional is too conservative and intensity is underprescribed, the client may drop out of the program due to lack of improvement and boredom. Therefore, to challenge your client effectively, an individualized approach to prescribing intensity is warranted.

To begin, the primary factor to consider when setting intensity is the client's initial fitness level at the start of the exercise program. This makes the evaluation of cardiorespiratory fitness a crucial piece for designing an aerobic exercise program. Variables such as goals, age, and health status, among others, should also be considered. To further help with pinpointing an appropriate intensity for clients, we will be employing the use of multiple ways to prescribe intensity. These methods are as follows.

HR Reserve and HRmax Methods

The most common way to prescribe intensity is using HR. Because of its assumed close relationship with $\dot{V}O_2$, HR is used as a predictor of $\dot{V}O_2$ responses during exercise. Because many

other factors (e.g., stress, anxiety, environmental temperature and humidity, medications, altitude, etc.) besides exercise can affect HR responses, it is prudent to pair a target HR (THR) recommendation with other intensity prescription methods to reduce the error associated with using HR alone (Gibson, Wagner, and Heyward 2019).

The HRmax method simply involves calculating a THR using percentages of measured maximal HR or predicted HRmax. The range of intensities to start an exercise program with that can be calculated using the HRmax method are from 55% to 63% (very light intensity for deconditioned clients) to 64% to 76% (moderate intensity for physically inactive/average fitness clients) and up to 77% to 90% (vigorous intensity for active clients to improve fitness) (ACSM 2021). Like we will see with other methods, it is typical to prescribe a range of intensities for the client. See the sample problem that follows.

Calculation example: 55-year-old client, age-predicted HRmax: 169 beats·min^{-1} (from Gellish et al. equation), THR range: 55% to 65% HRmax.

1. Calculate THR for low end of range.

THR = 169 beats·min^{-1} × 0.55 = 93 beats·min^{-1}

2. Calculate THR for high end of range.

THR = 169 beats·min^{-1} × 0.65 = 110 beats·min^{-1}

The HR reserve method involves calculating a THR from the HR reserve value. The range of intensities to start an exercise program with using the HR reserve method are from <40% (light intensity for deconditioned clients) to 40% to 59% (moderate intensity for physically inactive/average fitness clients) and up to 60% to 85% (vigorous intensity for active clients to improve fitness) (ACSM 2021). For this method, the HRmax can either be measured during a maximal exercise test or predicted from the Gellish et al. equation. Once HRmax is calculated, resting HR is subtracted from HRmax to get the HR reserve value. Intensity percentages are then taken from the HR reserve value and resting HR is added back in to get the THR. See the sample problem that follows.

Calculation example: 55-year-old client, age-predicted HRmax: 169 beats·min^{-1} (from Gellish et al. equation), resting HR: 69 beats·min^{-1}, THR range: 55% to 65% HR reserve.

1. Calculate HR reserve.

HR reserve = 169 beats·min^{-1} – 69 beats·min^{-1}
= 100 beats·min^{-1}

2. Calculate THR for low end of range.

THR = 100 beats·min^{-1} × 0.55
= 55 beats·min^{-1} + 69 beats·min^{-1}

THR = 124 beats·min^{-1}

3. Calculate THR for high end of range.

THR = 100 beats·min^{-1} × 0.65
= 65 beats·min^{-1} + 69 beats·min^{-1}

THR = 134 beats·min^{-1}

Comparing the two methods, the HR reserve method provides higher values than the HRmax method at similar relative percentages. Because this relationship holds true with fitness level and age, it is recommended to use the more conservative method of HRmax for older (>60 years old) and more sedentary individuals when prescribing intensity using HR (Kohrt et al. 1998). The advantage of using the HR reserve method is that it is more closely related to $\dot{V}O_2$ compared to the HRmax method, and is a better option for younger clients.

$\dot{V}O_2$ Reserve Method

Since $\dot{V}O_2$ is a good measure of exercise intensity and energy expenditure, it would be advantageous to use $\dot{V}O_2$ when prescribing exercise. The equation for this method is based on the HR reserve method where target intensities are taken from a $\dot{V}O_2$ reserve value. In this equation, $\dot{V}O_2$max is determined through maximal or submaximal exercise testing, and resting $\dot{V}O_2$ is 3.5 mL·kg^{-1}·min^{-1}. The range of intensities to start an exercise program using the $\dot{V}O_2$ reserve method are from <40% (light intensity for deconditioned clients) to 40% to 59% (moderate intensity for physically inactive/average fitness clients) and up to 60% to 85% (vigorous intensity for active clients to improve fitness) (ACSM 2021). Once $\dot{V}O_2$max is determined, resting $\dot{V}O_2$ is subtracted from $\dot{V}O_2$max to get the $\dot{V}O_2$ reserve value. Intensity percentages are then taken from the $\dot{V}O_2$ reserve value and resting $\dot{V}O_2$ is added back in to get the target $\dot{V}O_2$. See the sample problem that follows.

Calculation example: 55-year-old client, $\dot{V}O_2$max: 35 mL·kg^{-1}·min^{-1}, target $\dot{V}O_2$ range: 55% to 65% $\dot{V}O_2$ reserve.

1. Calculate $\dot{V}O_2$ reserve.

$\dot{V}O_2$ reserve = 35 mL·kg^{-1}·min^{-1} – 3.5 mL·kg^{-1}·min^{-1}
= 31.5 mL·kg^{-1}·min^{-1}

2. Calculate target $\dot{V}O_2$ for low end of range.

Target $\dot{V}O_2$ = 31.5 mL·kg^{-1}·min^{-1} × 0.55
= 17.3 mL·kg^{-1}·min^{-1} + 3.5 mL·kg^{-1}·min^{-1}

Target $\dot{V}O_2$ = 20.8 mL·kg^{-1}·min^{-1}

3. Calculate target $\dot{V}O_2$ for high end of range.

Target $\dot{V}O_2$ = 31.5 mL·kg^{-1}·min^{-1} × 0.65
= 20.5 mL·kg^{-1}·min^{-1} + 3.5 mL·kg^{-1}·min^{-1}

Target $\dot{V}O_2$ = 24 mL·kg^{-1}·min^{-1}

Once target $\dot{V}O_2$ values are calculated, they can be used to solve for workloads (e.g., speed, grade, work rate, etc.) using the ACSM metabolic equations. These values can also be converted to MET values (remember, divide by 3.5 mL·kg^{-1}·min^{-1}) and specific occupational, recreational, household, and so forth activities that match the MET values can be recommended to your client. See Ainsworth et al. (1993) for a comprehensive listing of those activities.

RPE Method

The concept of the RPE was first developed by Dr. Gunnar Borg in 1970 as a tool for gathering subjective feedback of exercise intensity based on perceived exertion (Borg 1982). Since that time, RPE has been regularly used and validated as an exercise intensity monitoring method in clinical, fitness, and athletic settings (Gibson, Wagner, and Heyward 2019). ACSM (2021) recommends a range of 11 to 16 on the 6 to 20 RPE scale for prescribing intensity. For less trained individuals, a light to moderate range of 11 to 13 is appropriate to start with in an exercise program, with a range of 13 to 16 for moderate to vigorous exercise either to progress to or start with if the client has a higher level of fitness. As an intensity monitoring method, RPE pairs well with HR, and exercise workload can be adjusted based on the HR response and how that corresponds with RPE in the exercise session. Some limitations of the RPE method are that it may take time and practice to use it effectively and some clients tend to both over- and underestimate their intensity. Also, RPE values can be mode specific; thus, consider that perceived intensity and exertion may be different from mode to mode in exercising clients. With this in mind, the RPE level should be determined by the individual mode (Zeni, Hoffman, and Clifford 1996).

Talk Test Method

The talk test is a method that is designed to guide perceived intensity based on how comfortably one can talk during exercise (Gibson, Wagner, and Heyward 2019). This method reflects the overall physiological relationship between exercise intensity and ventilation, meaning that as exercise intensity increases, so does ventilation and the difficulty of holding a comfortable conversation. The point at which individuals first have difficulty talking is associated with the ventilatory threshold (VT), which is a physiological marker for when steady-state exercise is no longer sustainable (Dehart-Beverly et al. 2000). Prescribing exercise using a threshold event like VT is much more individualized compared to the other methods described in this section because VT is related to certain metabolic changes (i.e., increased breathing, difficulty talking, greater fatigue, increased acidosis and blood lactate accumulation, etc.) that are common to all individuals (Porcari et al. 2018; Weatherwax et al. 2019). To use this method, have clients exercise while being mindful of their breathing and ability to talk. If their ability to talk is rated as comfortable, then they are exercising at appropriate intensities below VT. At the point when their talking becomes difficult, identify a HR and RPE specific to this intensity. From here, your client will know approximately what HR and RPE not to exceed so they can sustain steady-state exercise. Thus, intensity will be individualized to their own metabolic characteristics and guided by very clear physiological markers associated with their breathing.

Time (Duration)

This variable refers to the total amount of time that exercise is performed in one or multiple sessions per day. ACSM (2021) recommends a range of 30 to 60 $min \cdot d^{-1}$ for moderate intensity exercise and 20 to 60 $min \cdot d^{-1}$ for vigorous intensity exercise. Whatever time goal is chosen, clients can accumulate time throughout the day to reach the total time that is prescribed. In general, there is an inverse relationship between time and intensity, especially at the beginning of an exercise program (Gibson, Wagner, and Heyward 2019). Time will

usually be progressed more quickly relative to the progression of intensity over the duration of the training program.

Type (Mode)

Type or mode of exercise refers to the specific exercise(s) that your client will use in the aerobic exercise program. An activity is considered "aerobic" if it involves the use of large muscle groups, can be performed continuously, and is rhythmic in nature (Garber et al. 2011; Gibson, Wagner, and Heyward 2019). As stated earlier, it is advantageous for a client to perform a wide range of activities to promote variety of movement skill and muscle groups used and to lessen the chance of overtraining and overuse injuries (ACSM 2021; Garber et al. 2011). However, no matter what modality is used, the client needs to choose the mode(s) based on what they prefer and specific to their individual goals to maximize adherence to the exercise program. Other factors that can influence the choice of exercise are initial level of conditioning, prerequisite skill to perform the activity, functional limitations and need for activity modification, equipment availability, and special needs of the client (i.e., weight bearing versus nonweight bearing exercise).

Generally speaking, aerobic exercises can be grouped by modes that require minimal skill and where intensity can be controlled at low to moderate intensities (e.g., walking, cycling, etc.) and also vigorous intensities (e.g., jogging, running, rowing, etc.) (ACSM 2021). Furthermore, modes can be grouped by activities that require higher amounts of skill to perform (e.g., swimming, outdoor Nordic skiing, rollerblading, etc.) and activities that require greater skill and where intensity level is variable (e.g., basketball, tennis, soccer, etc.). Again, those clients who elect to engage in these last two groups of activities should have a regular exercise program in place and be of at least average cardiorespiratory fitness.

Volume

Exercise volume is determined by the combination of frequency, intensity, and time. It is usually represented as the total amount of exercise in time that is completed per week. ACSM (2021) recommends individuals accumulate at minimum 150 $min \cdot wk^{-1}$

for moderate intensity or 75 min·wk^{-1} of vigorous intensity exercise. If your client chooses, it has been shown that higher volumes of exercise up to 300 min·wk^{-1} for moderate and 150 min·wk^{-1} for vigorous are associated with greater health and fitness benefits (Bull et al. 2020; Garber et al. 2011; USDHHS 2018). In addition to minutes of exercise per week, there are other ways to quantify and prescribe volume: MET-min·wk^{-1}, energy expenditure, and step counts.

MET-min·wk^{-1}

The recommended target volume for MET-min·wk^{-1} is between 500 and 1,000 (ACSM 2021; Garber et al. 2011). The calculation of MET-min·wk^{-1} is a way to standardize exercise volume across varying modes and intensities to know if the program is meeting overall recommendations for activity level. This is particularly helpful in programs where a combination of both moderate and vigorous intensity exercise is prescribed. For example, a client is running 2 d·wk^{-1} for 30 min·d^{-1} at 8 METs (vigorous intensity) and walking 2 d·wk^{-1} for 45 min·d^{-1} at 4 METs (moderate intensity). Separately, 2 d of moderate and 2 d of vigorous exercise do not meet exercise recommendations, but combined over 4 d, the total activity might be enough. Here is how we can calculate MET-min·wk^{-1} to know:

1. Calculate MET-min·wk^{-1} for vigorous exercise.

$$\text{MET-min·wk}^{-1} = 2 \text{ d·wk}^{-1} \times 30 \text{ min·d}^{-1} \times 8 \text{ METs}$$

$$\text{MET-min·wk}^{-1} = 480 \text{ MET-min·wk}^{-1}$$

2. Calculate MET-min·wk^{-1} for moderate exercise.

$$\text{MET-min·wk}^{-1} = 2 \text{ d·wk}^{-1} \times 45 \text{ min·d}^{-1} \times 4 \text{ METs}$$

$$\text{MET-min·wk}^{-1} = 360 \text{ MET-min·wk}^{-1}$$

3. Calculate total MET-min·wk^{-1}.

$$\text{MET-min·wk}^{-1} = 480 \text{ MET-min·wk}^{-1} + 360 \text{ MET-min·wk}^{-1}$$

$$\text{MET-min·wk}^{-1} = 840 \text{ MET-min·wk}^{-1}$$

The total calculated MET-min·wk^{-1} falls within the recommended range of 500 to 1,000. With this value, we know that the combined moderate and vigorous intensity program is sufficient to meet recommendations for activity level.

Energy Expenditure

In lab 6, we learned how to calculate energy expenditure to determine how many calories are burned during exercise. Energy expenditure is an effective way to quantify exercise volume across varying modes and intensities. A general recommendation for adults to achieve during exercise is ≥1,000 kcal·wk^{-1} and ≥150 kcal per exercise session (ACSM 2021; Garber et al. 2011). This may be a difficult target for some to achieve, especially for very deconditioned individuals or beginners who are just starting a program. Thus, this can be an energy expenditure target to work toward and exceed as the exercise program progresses.

Another way to quantify exercise volume is through the calculation of energy expenditure using kcal·kg^{-1}·wk^{-1}. In this manner, energy expenditure is made relative or standardized to the client's body weight because weight can have influence over energy expenditure. According to Kraus et al. (2001), 14 kcal·kg^{-1}·wk^{-1} is equal to a moderate exercise volume and 23 kcal·kg^{-1}·wk^{-1} is equal to a high amount of exercise volume. To calculate your client's weekly energy expenditure goal, multiply body weight in kg by any number that falls within the range of 14 to 23 kcal·kg^{-1}·wk^{-1} for the desired exercise volume.

Step Counts

A final, easy method to quantify volume is through step counts. Current research recommends a daily step count in the range of 7,000 to 8,000 steps·d^{-1} at a rate of 100 steps·min^{-1} to achieve a moderate amount and intensity of exercise (Tudor-Locke et al. 2011; Tudor-Locke et al. 2018). With the widespread development of both simple (pedometers) and more advanced (accelerometers) wearable technology to gather this information, this could be another effective way to prescribe exercise for some clients.

Progression and Development of the Progression Plan

For most clients, a gradual progression of frequency, intensity, time, type of exercise, and volume over time is appropriate to reduce the risk of overtraining and injury plus promote a positive behavior change and adherence to the program (ACSM 2021). The rate at which to progress a client in a program depends on many factors, including current health status and cardiorespiratory fitness, how well they tolerate exercise, how fast they are adapting to the program, and the overall goals of the client. With these and other factors in mind, it is imperative that the exercise professional individualize the progression strategy to their specific client.

The plan outlined in the aerobic exercise programming progression chart involves three phases: the initial phase, the improvement phase, and the maintenance phase. The following are general recommendations for each phase in order to simplify your understanding of program design. Thus, design your program with these recommendations as your guide while keeping your client's specific goals and needs in mind.

Initial Phase

A main goal of the initial phase is to get your client ready for the increased stress of the improvement phase. Another goal is to promote adherence and enjoyment in exercise, so your client continues with the program. A more conservative approach is warranted in this phase, with the focus on starting at an appropriate intensity based on your client's initial fitness level and progressing to moderate intensity by the end of the phase (if appropriate).

Exercising at 3 d·wk^{-1} with a gradual increase of 5 to 10 min for overall time per session every 1 to 2 wk is a good target for the first 4 to 6 wk of the program (ACSM 2021; Garber et al. 2011). The length of the phase is dependent on the client and how long it takes to reach their initial goals.

Improvement Phase

The main goal of this phase is to gradually increase the exercise stimulus to elicit improvements in cardiorespiratory fitness over time. Depending on the client and individual goals, this phase can be as long as needed, but typically 3 mo or more. Again, a conservative approach is required here with a general recommendation of increasing only one FITT component at a time, increasing intensity by 5% every 1 to 2 wk if warranted, and increasing time 10% to 20% every 1 to 2 wk if warranted.

Maintenance Phase

The main goal of this phase is to maintain the gains realized in the improvement phase. Typically, program variables are held steady until a new program is started or adjusted in a way to maintain fitness improvements (i.e., drop volume and maintain or increase intensity).

Case Study for Aerobic Exercise Prescription

Looking at the aerobic exercise program progression chart, you will see that the first 2 wk of the progression plan are filled in to help you get started. Fill in the rest of the chart with your recommendations and calculated values for the case study client. With a real client, you would have the opportunity for a back-and-forth conversation and communication as the program is developed and progressed. However, do your best to apply these aerobic program design principles through this case study example and hand in your work to get feedback on your overall plan.

Demography

Age: 40

Height: 5 ft 6 in. (167.6 cm)

Weight: 155 lb (70.5 kg)

Sex: Female

Race or Ethnicity: Black or African American

Family History

Your client's family history reveals no cardiovascular issues.

Medical History

Present Conditions

On their health history questionnaire form, your client states no issues during exercise and various daily activities. A recent physical assessment revealed the following health information. Your client's resting HR was 70 beats·min^{-1} and their resting BP was 116/72 mmHg (confirmed from a previous measurement). The client's skinfold assessment showed a percent body fat (%BF) of 24%. No blood chemistry panel for cholesterol or blood glucose is reported at this time. Their last exercise test (6 mo ago), which was a submaximal walking test, showed a $\dot{V}O_2$max value of 30 mL·kg^{-1}·min^{-1}. No other data are available currently.

Past Conditions

Other than a minor surgical procedure on their shoulder 5 yr ago, your client reports no previous medical and health issues.

Behavior and Risk Assessment

Your client is a nonsmoker who has been physically active most of their life but is transitioning to a busier job that includes long periods of seated work. Your client is coming to you to develop a more organized and structured aerobic exercise program to increase cardiorespiratory fitness and learn about ways to limit sedentary behavior. Currently, they engage in 15 to 20 min of moderate intensity physical activity 3 d·wk^{-1} doing walking on a treadmill and 15 to 20 min of moderate intensity indoor cycling on the same days, which is a significant drop from what they used to do 1 yr ago. They would like to increase to 5 d·wk^{-1} and try to get back into jogging at a moderate to vigorous intensity (6 mi·hr^{-1} or 10 min·mi^{-1} pace) if possible by the end of the program. Overall, they are open to your suggestions. Your client's diet is good; meals are prepared at home with limited eating outside of the home. Please fill out the progression chart with the data for your client to follow based on their exercise goals.

CARDIORESPIRATORY FITNESS DATA SHEET

Name _____ Sex _____ Date _____

Age _____ Height _____ cm　Weight _____ kg

Resting HR _____ beats·min^{-1}　Resting BP _____ / _____ mmHg

Age-predicted HRmax _____ beats·min^{-1}

Gellish et al. formula: HRmax = 207 − (0.7 × age)

HR @ 70% HRR _____ beats·min^{-1}　HR @ 85% HRmax _____ beats·min^{-1}

Cycle Ergometry Tests: Modified YMCA Cycle Ergometer and George Submaximal Cycle Ergometer

Modified YMCA Cycle Ergometer Test

Stage	Work rate	Time	HR	RPE	BP
1	150 kg·m·min^{-1} or 25 W	1:30 min	_____ beats·min^{-1}	–	–
		2:30 min	_____ beats·min^{-1}	–	–
		2:45 min	–		_____ / _____ mmHg
2	_____ kg·m·min^{-1} or _____ W	1:30 min	_____ beats·min^{-1}	–	–
		2:30 min	_____ beats·min^{-1}	–	–
		2:45 min	–		_____ / _____ mmHg
3	_____ kg·m·min^{-1} or _____ W	1:30 min	_____ beats·min^{-1}	–	–
		2:30 min	_____ beats·min^{-1}	–	–
		2:45 min	–		_____ / _____ mmHg
4 (if needed)	_____ kg·m·min^{-1} or _____ W	1:30 min	_____ beats·min^{-1}	–	–
		2:30 min	_____ beats·min^{-1}	–	–
		2:45 min	–		_____ / _____ mmHg

Cool-down: 4-5 min at ≤150 kg·m·min^{-1} or 25 W, monitor HR and BP every 1-2 min

Note: Pedaling rate should be set at a constant 50 rev·min^{-1} to achieve the work rates in the table. Steady-state HR is defined as two consecutive HRs within 5 beats·min^{-1}. The goal is to obtain HRs between 110 beats·min^{-1} and 85% HRmax for at least two consecutive stages.

$\dot{V}O_2$max Calculation Variables From the Modified YMCA Exercise Test

$\dot{V}O_2$max = $\dot{V}O_2$SM$_2$ + [*b* (HRmax − HR$_2$)]
b = ($\dot{V}O_2$SM$_2$ − $\dot{V}O_2$SM$_1$) / (HR$_2$ − HR$_1$)

Stage	Work rate	$\dot{V}O_2$	HR
Final stage	_____ kg·m·min^{-1}	_____ mL·kg^{-1}·min^{-1} ($\dot{V}O_2$SM$_2$)	_____ beats·min^{-1} (HR$_2$)
Second to last stage	_____ kg·m·min^{-1}	_____ mL·kg^{-1}·min^{-1} ($\dot{V}O_2$SM$_1$)	_____ beats·min^{-1} (HR$_1$)
$\dot{V}O_2$max	_____ mL·kg^{-1}·min^{-1}		

Note: Remember, use the last two stages of the test to calculate $\dot{V}O_2$max. To calculate both $\dot{V}O_2$SM$_2$ and $\dot{V}O_2$SM$_1$, use the ACSM cycle ergometer metabolic equation from lab 6. You will need the work rate from both stages and body weight in kg.

George Submaximal Cycle Ergometer Test

$$\dot{V}O_2max = 85.447 + 9.104(sex) - 0.2676(age) - 0.4150(\text{body weight in kg}) + 0.1317(watts) - 0.1615(HR)$$

Stage	Work rate	Time	HR		RPE
Warm-up	420 kg·m·min^{-1} or ~68 W	2:00 min	_____	beats·min^{-1}	
1	_____ kg·m·min^{-1} or _____ W	2:00 min	_____	beats·min^{-1}	–
		3:00 min	_____	beats·min^{-1}	
2 (if needed)	_____ kg·m·min^{-1} or _____ W	2:00 min	_____	beats·min^{-1}	–
		3:00 min	_____	beats·min^{-1}	
Cool-down: 3-5 min at ≤150 kg·m·min^{-1} or 25 W, monitor HR and BP every 1-2 min					
$\dot{V}O_2max$	_____ mL·kg^{-1}·min^{-1}				

Note: Pedaling rate should be set at a constant 70 rev·min^{-1} to achieve the work rates in the table. Steady-state HR is defined as two consecutive HRs within 6 beats·min^{-1}. The HR in the last stage of the test (either stage 1 or 2) should be between 120 and 175 beats·min^{-1}. Continue this process until a steady-state HR between the goal range is elicited. For sex, 1 will be entered for men and 0 will be entered for women.

Treadmill Exercise Tests: Ebbeling Walk and George Jog

Ebbeling Walking Test

$$\dot{V}O_2max = 15.1 + 21.8(\text{speed in mi·h}^{-1}) - 0.327(HR) - 0.263(\text{speed} \times age) + 0.00504(HR \times age) + 5.48(sex)$$

Walking speed during 4 min warm-up (0% grade) Warm-up HR (should be 50%-70% HRmax)	_____ mi·h^{-1} _____ beats·min^{-1}
HR at selected walking speed at 5% grade	_____ beats·min^{-1} _____ beats·min^{-1}
$\dot{V}O_2max$	_____ mL·kg^{-1}·min^{-1}

Note: The walking pace established in the warm-up period should be brisk (2.0-4.5 mi·h^{-1} or 53.6-120.6 m·min^{-1}). Measure HR near the end of the fourth minute during the walking stage at 5% grade. Record two HRs during this time to establish steady-state agreement (within 5-6 beats·min^{-1}). For sex, 1 will be entered for men and 0 will be entered for women. Follow general guidelines for cool-down (at least 4-5 min).

George Jog Test

$$\dot{V}O_2max = 54.07 - 0.1938(\text{weight in kg}) + 4.47(\text{speed in mi·h}^{-1}) - 0.1453(HR) + 7.062(sex)$$

Jogging speed during exercise test (0% grade)	_____ mi·h^{-1}
HR following 3 min of jogging at a constant pace	_____ beats·min^{-1} _____ beats·min^{-1}
$\dot{V}O_2max$	_____ mL·kg^{-1}·min^{-1}

Note: The jogging pace established can be adjusted until comfortable for the client (4.3-7.5 mi·h^{-1} or 115.2-201 m·min^{-1}). Women should not exceed 6.5 mi·h^{-1} (174.2 m·min^{-1}) and men should not exceed 7.5 mi·h^{-1} (201 m·min^{-1}) during the test. Measure HR following 3 min of jogging. Record two consecutive HRs (30 s apart) until steady-state agreement (≤3 beats·min^{-1}) is established. HR should not exceed 180 beats·min^{-1}. For sex, 1 will be entered for men and 0 will be entered for women. Follow general guidelines for cool-down (at least 4-5 min).

> continued

Cardiorespiratory Fitness Data Sheet   > *continued*

Field Exercise Tests: Rockport 1-Mile Walk, George 1-Mile Run, and Larsen 1.5-Mile Run/Walk

Rockport 1-Mile (1.6 km) Walk Test

$\dot{V}O_2max = 132.853 - 0.0769(\text{weight in kg}) - 0.3877(\text{age}) + 6.315(\text{sex}) - 3.2649(\text{time in min}) - 0.1565(\text{HR})$

Walking time in minutes	_____ min
HR at completion of the test	_____ beats·min^{-1}
$\dot{V}O_2max$	_____ mL·kg^{-1}·min^{-1}

Note: Convert time in minutes and seconds to minutes only. The client should be encouraged to walk as quickly as possible during the test. Measure HR via radial pulse immediately after the test is completed using a 15 s count (multiply by 4). If measuring with an electronic device, measure during this time period. A 5-10 min warm-up prior to the test is recommended. For sex, 1 will be entered for men and 0 will be entered for women. Follow general guidelines for cool-down (at least 4-5 min).

George 1-Mile (1.6 km) Run Test

$\dot{V}O_2max = 100.5 - 0.1636(\text{weight in kg}) - 1.438(\text{time in min}) - 0.1928(\text{HR}) + 8.344(\text{sex})$

Running time in minutes	_____ min
HR at completion of the test	_____ beats·min^{-1}
RPE at completion of the test	_____ RPE
$\dot{V}O_2max$	_____ mL·kg^{-1}·min^{-1}

Note: Convert time in minutes and seconds to minutes only. The client should be encouraged to run at a moderate-intensity pace during the test and time of completion should be ≥8 min for men and ≥9 min for women. Measure HR via radial pulse immediately after the test is completed using a 15 s count (multiply by 4). If measuring with an electronic device, measure during this time period. HR should not exceed 180 beats·min^{-1}. A 2-3 min warm-up prior to the test is recommended. For sex, 1 will be entered for men and 0 will be entered for women. Follow general guidelines for cool-down (at least 4-5 min).

Larsen 1.5-Mile (2.4 km) Run/Walk Test

$\dot{V}O_2max = 100.162 + 7.301(\text{sex}) - 0.164(\text{weight in kg}) - 1.273(\text{time in min}) - 0.156(\text{HR})$

Running time in minutes	_____ min
Target HR range during test	_____ beats·min^{-1}
Target RPE during the test	_____ RPE
HR at completion of the test	_____ beats·min^{-1}
$\dot{V}O_2max$	_____ mL·kg^{-1}·min^{-1}

Note: Convert time in minutes and seconds to minutes only. The client should be encouraged to run or walk at a steady pace equivalent to a "somewhat hard" (RPE = 13) intensity during the test. Clients should maintain their exercise test HR between 60%-90% HRmax. Measure HR at the completion of the test either via palpation or electronic device. For sex, 1 will be entered for men and 0 will be entered for women. Follow general guidelines for cool-down (at least 4-5 min).

From J. Janot and N. Beltz, *Laboratory Assessment and Exercise Prescription* (Champaign, IL: Human Kinetics, 2023).

AEROBIC EXERCISE PROGRAMMING PROGRESSION CHART

INITIAL PHASE: WEEKS 1-2 (FREQUENCY: 3 D·WK⁻¹ AND TIME: 60 MIN·D⁻¹ [30 MIN FOR EACH MODALITY])

Type	%$\dot{V}O_2R$	$\dot{V}O_2$ (mL·kg⁻¹·min⁻¹)	%HRR	HR	METs	MET-min·wk⁻¹	RPE	kcal·d⁻¹ (session)	kcal·wk⁻¹	kcal·kg⁻¹·wk⁻¹	Speed (mi·h⁻¹)	Pace (min·mi⁻¹)
Walking	50%-55%	16.8-18.1	50%-55%	125-130	4.8-5.1	432-459	12-13	141-155	423-465	6-6.6	2.9-3.2 @ 4% grade	20:40-18:45 (flat ground)

Type	%$\dot{V}O_2R$	$\dot{V}O_2$ (mL·kg⁻¹·min⁻¹)	%HRR	HR	METs	MET-min·wk⁻¹	RPE	kcal·d⁻¹ (session)	kcal·wk⁻¹	kcal·kg⁻¹·wk⁻¹	Workload (W)	
Cycling	55%-60%	18.1-19.4	55%-60%	130-135	5.1-5.5	459-495	12-13	155-168	465-504	6.6-7.1	71-79	

INITIAL PHASE: WEEKS 3-4 (FREQUENCY _____ AND TIME _____)

Type	%$\dot{V}O_2R$	$\dot{V}O_2$	%HRR	HR	METs	MET-min·wk⁻¹	RPE	kcal·d⁻¹ (session)	kcal·wk⁻¹	kcal·kg⁻¹·wk⁻¹	Speed (mi·h⁻¹)	Pace (min·mi⁻¹)

Type	%$\dot{V}O_2R$	$\dot{V}O_2$	%HRR	HR	METs	MET-min·wk⁻¹	RPE	kcal·d⁻¹ (session)	kcal·wk⁻¹	kcal·kg⁻¹·wk⁻¹	Workload (W)	

IMPROVEMENT PHASE: WEEKS 5-8 (FREQUENCY _____ AND TIME _____)

Type	%$\dot{V}O_2R$	$\dot{V}O_2$	%HRR	HR	METs	MET-min·wk⁻¹	RPE	kcal·d⁻¹ (session)	kcal·wk⁻¹	kcal·kg⁻¹·wk⁻¹	Speed (mi·h⁻¹)	Pace (min·mi⁻¹)

Type	%$\dot{V}O_2R$	$\dot{V}O_2$	%HRR	HR	METs	MET-min·wk⁻¹	RPE	kcal·d⁻¹ (session)	kcal·wk⁻¹	kcal·kg⁻¹·wk⁻¹	Workload (W)	

IMPROVEMENT PHASE: WEEKS 9-12 (FREQUENCY _____ AND TIME _____)

Type	%$\dot{V}O_2R$	$\dot{V}O_2$	%HRR	HR	METs	MET-min·wk⁻¹	RPE	kcal·d⁻¹ (session)	kcal·wk⁻¹	kcal·kg⁻¹·wk⁻¹	Speed (mi·h⁻¹)	Pace (min·mi⁻¹)

Type	%$\dot{V}O_2R$	$\dot{V}O_2$	%HRR	HR	METs	MET-min·wk⁻¹	RPE	kcal·d⁻¹ (session)	kcal·wk⁻¹	kcal·kg⁻¹·wk⁻¹	Workload (W)	

MAINTENANCE PHASE: WEEKS 13+ (FREQUENCY _____ AND TIME _____)

Type	%$\dot{V}O_2R$	$\dot{V}O_2$	%HRR	HR	METs	MET-min·wk⁻¹	RPE	kcal·d⁻¹ (session)	kcal·wk⁻¹	kcal·kg⁻¹·wk⁻¹	Speed (mi·h⁻¹)	Pace (min·mi⁻¹)

Type	%$\dot{V}O_2R$	$\dot{V}O_2$	%HRR	HR	METs	MET-min·wk⁻¹	RPE	kcal·d⁻¹ (session)	kcal·wk⁻¹	kcal·kg⁻¹·wk⁻¹	Workload (W)	

From J. Janot and N. Beltz, *Laboratory Assessment and Exercise Prescription* (Champaign, IL: Human Kinetics, 2023).

Assessment of Aerobic Fitness: Maximal Oxygen Consumption

Purpose of the Lab

As you learned in lab 7, the estimation of maximal oxygen consumption ($\dot{V}O_2max$) is useful in a variety of scenarios. Measuring $\dot{V}O_2max$ via indirect calorimetry is considered the gold standard for determining an individual's cardiorespiratory fitness (CRF) level. The process of quantifying $\dot{V}O_2max$ is done through the systematic application of incremental exercise intensity during a graded exercise test (GXT). Generally, exercise intensity, or workload, begins well below maximal levels during a GXT. Intensity is added gradually throughout the test until the participant reaches volitional exhaustion or is no longer able to tolerate the workload. The fashion in which the workload is added depends on the exercise modality and overall design of the GXT.

The two most common GXT modalities are treadmill (figure 8.1) and cycle ergometer (figure 8.2), although a $\dot{V}O_2max$ test can be performed on any piece of exercise equipment that allows the test administrator to manipulate the workload

systematically. Treadmill is the most widely used GXT modality because individuals are familiar with upright locomotion, and it uses the greatest amount of muscle mass. Cycle ergometers can also be used if an individual has underlying orthopedic issues that preclude safe treadmill exercise or if the exerciser is more accustomed to regular exercise on a cycle compared to walking or running. Together with modality, the rate and fashion in which workload increases during the GXT stages can vary. The two most common incremental workload schemes are step and ramp stage designs. Step protocols can range between 1 and 3 min per stage during a GXT; the longer the stage duration, the greater the increase in workload between each stage. Ramp designs provide nearly imperceptible increases in workload throughout stages that last anywhere from 2 to 15 s in duration. A GXT should also aim for 8 to 12 min total test duration (Buchfuhrer et al. 1983). Studies have also turned to a new approach in GXT design, a self-paced GXT where the exerciser adjusts the workload in accordance with stages set by the 6 to 20 rating of perceived

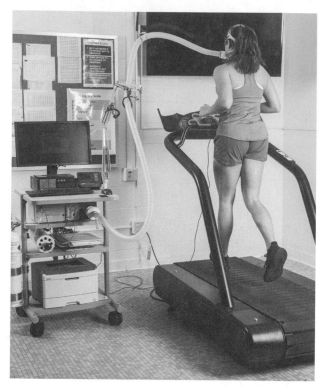

FIGURE 8.1 Metabolic cart with treadmill.

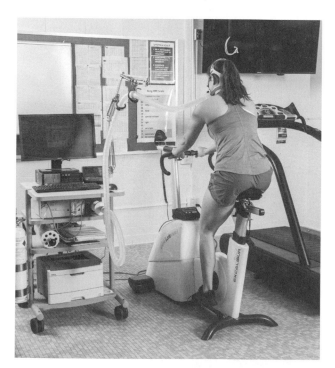

FIGURE 8.2 Electronically braked cycle ergometer.

exertion scale (RPE$_{6-20}$) (Borg 1982; Eston et al. 2005; Mauger and Sculthorpe 2012). The details of these GXT designs will be explained in greater detail throughout this lab.

It is important to understand that the foundational basis of $\dot{V}O_2$max testing is that it represents the uppermost limit of an individual's ability to do work. It is common for an individual's treadmill $\dot{V}O_2$max to be up to 20% higher compared to their $\dot{V}O_2$max on a cycle ergometer (Myers et al. 1991). For this reason, we cannot conclude that both tests represent that individual's $\dot{V}O_2$max when one is clearly greater than the other despite the individual reaching volitional exhaustion during both tests. It is common to use the term $\dot{V}O_2$peak during these scenarios. A $\dot{V}O_2$peak represents the highest $\dot{V}O_2$ achieved during a GXT but does not necessarily represent the true physiological ceiling for that individual. In other words, a $\dot{V}O_2$max is always a $\dot{V}O_2$peak but a $\dot{V}O_2$peak is never a $\dot{V}O_2$max. There are many reasons why only a $\dot{V}O_2$peak may be reached during a test, such as an individual giving less than maximal effort, localized muscular fatigue rather than global exhaustion, less muscle mass utilized during a particular modality, orthopedic issues, or lack of familiarization with the modality.

A set of criteria has been established to verify maximal effort and the attainment of $\dot{V}O_2$max based on a plateau in $\dot{V}O_2$ response, proximity to estimated age-predicted heart rate maximum (HRmax), maximal respiratory exchange ratio (RERmax), reported RPE$_{6-20}$, and blood lactate accumulation (BLa$^-$) (Beltz et al. 2016). A viable alternative to the RPE$_{6-20}$ is a category scale with ratio properties on a 0 to 10 scale (CR10) (Borg 1982). The CR10 is a viable alternative to the 6 to 20 scale to indicate perceived effort. The 0 to 10 scale can be more easily interpreted by some individuals compared to the 6 to 20 scale. The most common thresholds for the $\dot{V}O_2$max attainment criteria are as follows:

- $\dot{V}O_2$ plateau of $\leq$150 mL·min^{-1}
- HRmax within 10 beats·min^{-1} of age-predicted HRmax
- RERmax $\geq$1.15
- RPE$_{6-20}$ $\geq$17 or CR10 $\geq$8
- BLa$^-$ $\geq$8.0 mM

Please note age-predicted HRmax is calculated using the Gellish et al. (2007) equation:

$$\text{Age-Predicted HRmax} = 207 - (0.7 \times \text{age})$$

Interestingly, the application of these attainment criteria is not standardized. Which and how many of these individual criteria are used is determined by the preference of the test administrator. For this reason, the primary criterion to confirm attainment of $\dot{V}O_2$max is a $\dot{V}O_2$ plateau ≤150 mL·min^{-1}. If the testing data show this plateau in oxygen consumption despite increasing muscular workload, it is verified as a true $\dot{V}O_2$max (Bassett and Howley 2000). The HRmax, RERmax, RPE$_{6-20}$, and BLa$^-$ thresholds can be thought of as secondary criteria. Any combination of these can be used given the capabilities of the exercise laboratory and preference of the test administrator. For example, BLa$^-$ testing requires additional equipment, and the collection of blood droplets may not even be allowed in certain circumstances. For this reason, we will not be discussing the methods for BLa$^-$ collection in this lab.

Another useful application of metabolic gas data is ventilatory threshold (VT) identification. As workload increases, a point is reached where metabolic disturbances within the exercising skeletal muscle cause hydrogen ion accumulation beyond the capabilities of bicarbonate buffering in the bloodstream. The result is an excess of CO_2 production and sharp increases in VT. This can occur twice during incremental exercise, and these points are referred to as the first (VT1) and second (VT2) ventilatory thresholds. Metabolic data analysis shows that VT1 occurs at a point where $\dot{V}E$ increases disproportionately to $\dot{V}O_2$ while VT2 occurs at the point where $\dot{V}E$ increases disproportionately to $\dot{V}CO_2$ (Beaver, Wasserman, and Whipp 1986). These thresholds can be used to design an individualized exercise prescription that may be favorable over standard heart rate reserve (HRR) or $\dot{V}O_2$ reserve ($\dot{V}O_2$R) methods (Weatherwax et al. 2019).

The primary purpose of this lab is to investigate multiple GXT protocols and analyze the exercise data to (1) verify the attainment of $\dot{V}O_2$max, (2) compare $\dot{V}O_2$max values across modalities, and (3) identify VT for the purpose of exercise prescription.

Necessary Equipment or Materials

- Metabolic analyzer and GXT equipment (e.g., breathing tube, mouthpiece, air filter)
- Treadmill or electronically braked cycle ergometer
- HR monitor
- Body weight scale
- RPE$_{6-20}$ or CR10 scale
- Spreadsheet computer software
- Maximal oxygen consumption data sheet

Calibration of Equipment

Calibration of the metabolic analyzer should be performed in accordance with the manufacturer guidelines. Generally, indirect calorimetry systems need to warm up for at least 30 min prior to calibration. They are then calibrated through a series of gas and flow meter calibration steps. Some of the modalities used in this lab, such as motorized treadmills and electronically braked cycle ergometers, cannot be calibrated within the exercise physiology lab setting. In the case where these devices are not working properly, the manufacturer would have to service them to restore their accuracy.

Procedures

Pretest Instructions

1. Medical clearance for testing is recommended for individuals with known metabolic, renal, or cardiovascular disease. Once clearance is obtained, a healthcare provider should be present for maximal exercise testing.
2. The participant should arrive adequately hydrated and refrain from eating a heavy meal within 2 h prior to testing.

Testing

1. Prior to the participant arriving, construct a mask/mouthpiece, attach a breathing tube to the analyzer, and place air filters in accordance with the system manufacturer's guidelines.
2. Record the participant's body weight to the nearest 0.1 kg.
3. Allow the participant to warm up on the modality they will be using during the GXT (treadmill or cycle). They should exercise at a self-selected intensity of 11 ("light") using the RPE$_{6-20}$ scale or 2 to 3 ("slight" to "moderate") on the CR10 scale for approximately 3 to 5 min. Increase the workload on the treadmill via speed and/or grade and on the cycle ergometer via power output in watts (W).

4. Instruct the participant on the use of the RPE_{6-20} or CR10 scales. Explain the entire testing procedure and what they can expect during the GXT.

5. Secure a heart rate monitor on the participant at the level of the xiphoid process, under the shirt. To improve conductivity and signal, apply water or electrode gel to the surface of the monitor that is contacting the skin.

6. Attach the breathing hose to the participant's mask or mouthpiece.

Cycle Ramp Protocol

1. Adjust the seat to the appropriate height for the participant. Align the seat to the level of the greater trochanter and then make any adjustments to ensure a 5- to 10-degree bend in the knee at the bottom of the pedal revolution. Adjust handlebars to level of comfort.

2. Select a ramp protocol specific to the participant's level of training/fitness.

    Male:

    Trained cyclist: 60 W start, 30 $W \cdot min^{-1}$ ramp

    Active: 50 W start, 25 $W \cdot min^{-1}$ ramp

    Inactive: 40 W start, 20 $W \cdot min^{-1}$ ramp

    Female:

    Trained cyclist: 50 W start, 25 $W \cdot min^{-1}$ ramp

    Active: 40 W start, 20 $W \cdot min^{-1}$ ramp

    Inactive: 30 W start, 15 $W \cdot min^{-1}$ ramp

    The ramp test using an electronically braked cycle ergometer requires software that controls workload increments.

3. Encourage the participant to maintain a cadence between 70 and 90 rpm.

4. Participant must remain seated during the entire duration of the GXT.

5. Always be sure to monitor the participant for signs or symptoms of an adverse cardiovascular event. Review lab 7 for the indications for GXT termination.

6. Record HR, RER, RPE_{6-20} or CR10, and workload every 2 min during the exercise on the lab 8 data sheet.

7. GXT is terminated when:

    a. Participant is no longer able to maintain a cadence of >50 rpm or

    b. participant experiences signs or symptoms for GXT termination.

8. Use the highest 15 s average data point as $\dot{V}O_2$max.

9. Interpret the $\dot{V}O_2$max using tables 7.2 and 7.3.

Treadmill Step Protocol

1. The step GXT completed during this lab is the Bruce protocol (Bruce et al. 1963). It is one of the most widely used treadmill GXT protocols in clinical, research, and performance settings. The stages for the Bruce protocol are as follows:

    Stage 1 (0-3 min): 1.7 mph, 10% grade

    Stage 2 (3-6 min): 2.5 mph, 12% grade

    Stage 3 (6-9 min): 3.4 mph, 14% grade

    Stage 4 (9-12 min): 4.2 mph, 16% grade

    Stage 5 (12-15 min): 5.0 mph, 18% grade

    Stage 6 (15-18 min): 5.5 mph, 20% grade

2. Always be sure to monitor the participant for signs or symptoms of an adverse cardiovascular event. Review lab 7 for the indications for GXT termination.

3. Allow minimal handrail grasping from the participant. Handrail support should be used sparingly to assist in balance and coordination because it will blunt $\dot{V}O_2$ and HR responses during exercise (Berling et al. 2006).

4. Record HR, RER, RPE_{6-20} or CR10, and workload at the end of every stage on the lab 8 data sheet.

5. GXT is terminated when the following occurs:

    a. Participant reaches volitional exhaustion. *Tip:* Instruct the participant that volitional exhaustion is a global fatigue involving all the systems. A realistic cue for volitional exhaustion is if they cannot maintain the current workload for 15 to 30 more seconds, or

    b. participant experiences signs or symptoms for GXT termination.

6. Use the highest 15 s average data point as $\dot{V}O_2$max.

7. Interpret the $\dot{V}O_2$max using tables 7.2 and 7.3.

Self-Paced Protocol

1. The self-paced GXT in this lab can be completed on either a treadmill or an electronically braked cycle ergometer. The participant must have the ability to access the workload controls to adjust during testing.

2. This GXT protocol is a 10 min test consisting of five 2 min stages. Each stage corresponds to a specific value on the RPE_{6-20} scale. The participant is free to increase or decrease speed (treadmill) or power output (cycle) at any given moment during the test but must work at the designated RPE_{6-20} level. The test is performed at a constant 3% incline on a treadmill. The stages and RPE_{6-20} workloads for the self-paced GXT are:

   Stage 1 (0-2 min): 11 or "fairly light"

   Stage 2 (2-4 min): 13 or "somewhat hard"

   Stage 3 (4-6 min): 15 or "hard"

   Stage 4 (6-8 min): 17 or "very hard"

   Stage 5 (8-10 min): 20 or "maximal"

3. Encourage the participant during each stage to match their effort with the RPE_{6-20} workload.

4. Always be sure to monitor the participant for signs or symptoms of an adverse cardiovascular event. Review lab 7 for the indications for GXT termination.

5. Record HR, RER, RPE_{6-20} or CR10, and workload at the end of every stage on the lab 8 data sheet.

6. GXT is terminated when the following occurs:

   a. The 10 min exercise duration is completed or

   b. participant experiences signs or symptoms for GXT termination.

7. Use the highest 15 s average data point as $\dot{V}O_2$max.

8. Interpret the $\dot{V}O_2$max using tables 7.2 and 7.3.

$\dot{V}O_2$max Attainment Criteria

Primary

1. View the GXT data in 30 s sample average. If viewing the cycle ramp or self-paced data, compare the final two $\dot{V}O_2$ data points during exercise. If viewing step protocol data, compare the highest $\dot{V}O_2$ data points in the final two stages. Confirm a $\dot{V}O_2$ plateau if these two data points are ≤ 150 mL·min^{-1}.

Secondary

1. View the GXT data in single breath output. Compare the highest recorded HR on the output to the participant's age-predicted HRmax. Attainment criteria is met if exercise HRmax is within 10 beats·min^{-1} of age-predicted HRmax.

2. View the GXT data in 15 s sample average. Find the participant's highest recorded RER value. Attainment criteria is met if RERmax is ≥ 1.15.

3. Attainment criteria is met if highest exercise RPE_{6-20} ≥ 17 or CR10 ≥ 8.

Identifying Ventilatory Thresholds

Export the 15 s sample average GXT data to a spreadsheet computer software program. Clean up the data by removing all vertical columns of data except Time, $\dot{V}E$, $\dot{V}O_2$, and $\dot{V}CO_2$. You will need to create two separate figures to identify VT1 and VT2.

Ventilatory Threshold 1 (VT1)

Select all the exercise data and create a figure to display $\dot{V}E$ (L·min^{-1}) on the vertical axis and $\dot{V}O_2$ (L·min^{-1}) on the horizontal axis. Format the vertical and horizontal units to minimize the amount of empty space in the figure display area. You can immediately see that the response is curvilinear, containing two distinct portions and slopes. Using the line drawing tool, draw a line that best fits the initial slope of the data by starting with the lowest data points. Draw another best-fit line to represent the second, steep portion of the data set. Draw a line straight from the point of intersection down to the $\dot{V}O_2$ axis; this value represents VT1. Interpret the VT1 value in mL·kg^{-1}·min^{-1} using table 8.1 or 8.2.

TABLE 8.1   Ventilatory Threshold 1 (mL·kg⁻¹·min⁻¹) Percentiles From Treadmill Graded Exercise Testing by Age and Sex

Age (yr)	PERCENTILE						
	5	10	25	50	75	90	95
MEN							
20-29	13.8	15.6	17.9	22.0	26.3	30.4	34.4
30-39	14.0	15.1	17.8	21.3	26.2	30.6	33.3
40-49	12.9	14.4	16.7	20.1	24.1	28.8	32.6
50-59	12.6	13.8	15.6	18.2	22.8	27.3	30.4
60-69	12.4	12.5	15.0	17.8	21.0	24.3	28.0
70-79	10.0	11.5	13.4	16.5	22.1	24.5	25.6
WOMEN							
20-29	11.7	13.7	16.3	21.7	27.7	34.2	40.6
30-39	11.6	12.5	13.7	16.2	19.3	23.5	24.9
40-49	10.7	12.0	14.3	16.2	19.3	23.5	24.9
50-59	10.6	12.0	13.1	15.0	17.4	20.6	23.2
60-69	8.9	10.4	12.1	13.5	16.0	18.8	19.7
70-79	7.0	9.0	10.2	12.3	14.2	17.9	20.1

Reprinted by permission from B. Vainshelboim, R. Arena, L.A. Kaminsky, and J. Myers, "Reference Standards for Ventilatory Threshold Measured With Cardiopulmonary Exercise Testing. The Fitness Registry and the Importance of Exercise: A National Database," *CHEST* 157, no. 6 (2020): 1531-1537.

TABLE 8.2   Ventilatory Threshold 1 (mL·kg⁻¹·min⁻¹) Percentiles From Cycle Graded Exercise Testing by Age and Sex

Age (yr)	PERCENTILE						
	5	10	25	50	75	90	95
MEN							
20-29	12.0	13.0	15.0	17.8	20.9	23.4	25.8
30-39	11.0	12.0	14.0	16.0	19.0	22.0	24.0
40-49	10.0	11.0	13.0	15.0	17.3	20.0	22.0
50-59	10.0	10.4	12.0	14.0	16.0	18.9	20.0
60-69	9.1	10.0	11.0	13.0	15.0	17.0	19.1
70-79	8.0	8.7	10.0	11.0	13.1	15.0	16.1
WOMEN							
20-29	8.5	9.4	11.5	13.9	16.9	20.0	24.8
30-39	8.2	9.1	10.2	12.3	15.8	19.1	22.8
40-49	7.9	8.6	10.1	12.3	15.6	19.0	21.0
50-59	7.7	8.6	10.0	11.9	14.2	17.7	19.9
60-69	7.2	7.7	9.5	11.2	12.9	16.4	18.2
70-79	5.7	7.8	8.9	10.7	11.6	13.8	14.7

Reprinted by permission from B. Vainshelboim, R. Arena, L.A. Kaminsky, and J. Myers, "Reference Standards for Ventilatory Threshold Measured With Cardiopulmonary Exercise Testing. The Fitness Registry and the Importance of Exercise: A National Database," *CHEST* 157, no. 6 (2020): 1531-1537.

Ventilatory Threshold 2 (VT2)

Select all the exercise data and create a figure to display $\dot{V}E$ (L·min^{-1}) on the vertical axis and $\dot{V}CO_2$ (L·min^{-1}) on the horizontal axis. Format the vertical and horizontal units to minimize the amount of empty space in the figure display area. You can immediately see that the response is curvilinear, containing two distinct portions and slopes. Using the line drawing tool, draw a line that best fits the initial slope of the data by starting with the lowest data points. Draw another best-fit line to represent the second, steep portion of the data set. Draw a line straight from the point of intersection down to the $\dot{V}CO_2$ axis. Locate this $\dot{V}CO_2$ data point on the spreadsheet to identify the corresponding $\dot{V}O_2$ data point; this value represents VT2.

Exercise Program Design: Alternative Approach to Aerobic Prescription

An investigation by Weatherwax et al. (2019) proposed a solution to the issue of "nonresponders" to traditional $\dot{V}O_2R$ and HRR aerobic exercise prescription methods. Some individuals are prescribed and follow aerobic training programs, yet the program does not elicit positive changes to their CRF level. One of the underlying factors that explains the nonresponder is that there is considerable variability across individuals when it comes to important metabolic disturbances during exercise. These metabolic disturbances are better known at VT1 and VT2, both of which were identified in this lab activity. The results of

the Weatherwax et al. (2019) study showed that 60% of the traditional prescription (% HRR) group responded to the 12 wk training program (>4.7% increase in $\dot{V}O_2$max) while 100% of the threshold prescription responded to the training. Exercise prescription using VT1 and VT2 follows the same aerobic exercise principles that you learned in lab 7; however, we will set exercise intensity based on VT1 and VT2 identification. Following is a summary of a 12 wk aerobic intensity prescription using ventilatory thresholds.

Weeks 1 to 3: HR < VT1

Target HR < VT1 = Range from 10 beats·min^{-1} below VT1 to HR at VT1

Weeks 4 to 8: HR ≥ VT1 to ≤ VT2

Target HR ≥ VT1 to ≤ VT2 = HR range of 15 beats·min^{-1} directly between VT1 and VT2

Weeks 9 to 12: HR ≥ VT2

Target HR ≥ VT2 = Range from HR at VT2 to 10 beats·min^{-1} above VT2

For example, you complete a treadmill GXT on a 25-year-old male and collect the following data (table 8.3):

$\dot{V}O_2$max = 43.2 mL·kg^{-1}·min^{-1}

HRmax = 202 beats·min^{-1}

VT1 = 24.7 mL·kg^{-1}·min^{-1}

HR at VT1 = 114 beats·min^{-1}

VT2 = 33.6 mL·kg^{-1}·min^{-1}

HR at VT2 = 151 beats·min^{-1}

Refer to the guidelines in lab 7 to follow the frequency and volume components of the aerobic exercise prescription.

TABLE 8.3   12-Week Aerobic Intensity Prescription

Week	Target heart rate	HR range (beats·min^{-1})
1 to 3	HR < VT1	104-114 beats·min^{-1}
4 to 8	HR ≥ VT1 to ≤ VT2	125-140 beats·min^{-1}
9 to 12	HR ≥ VT2	151-161 beats·min^{-1}

MAXIMAL OXYGEN CONSUMPTION

Name _____ Date _____

Age _____ Height _____ cm Weight _____ kg ☐ Female ☐ Male

Age-Predicted HRmax: $207 - (0.7 \times age) =$ _____ beats·min^{-1}

Modality _____ Protocol _____

Stage	Time (min)	Speed (mph)	Grade (%)	Power output (W)	HR (bpm)	RPE	RER
1							
2							
3							
4							
5							
6							
7							
8							

Attainment Criteria

Primary

$\dot{V}O_2$max _____ mL·kg^{-1}·min^{-1}

☐ $\dot{V}O_2$ plateau of ≤150 mL·min^{-1}

Evaluation _____

Secondary

VT1 _____ mL·kg^{-1}·min^{-1}

VT2 _____ mL·kg^{-1}·min^{-1}

☐ HRmax within 10 beats·min^{-1} of age-predicted HRmax

☐ RERmax ≥1.15

☐ RPE$_{6-20}$ ≥17 or CR10 ≥8

Assessment of Muscular Fitness: Flexibility and Range of Motion

Purpose of the Lab

The evaluation of flexibility and range of motion (ROM) in a comprehensive health and fitness assessment is done for several reasons. First, it is important to establish baseline flexibility and ROM in a needs analysis so an exercise professional can design an individualized program to target areas of the body that are tight or movement restricted. This can then be tracked over time to determine the overall effectiveness of the training program. It is also important to assess changes in flexibility across the lifespan so that interventions to maintain independence and the ability to perform activities of daily living can be implemented. Additionally, the assessment of flexibility in a competitive athlete can be used to determine if either prerequisite ROM needs to perform the sport have been met or if stabilization of hypermobile joints to prevent injury is more warranted than enhancing ROM. Therefore, the assessment of joint ROM and muscular flexibility is a crucial piece to complete the picture of your client's overall health and physical function.

Flexibility can be defined as the ability of a joint or series of joints to move fluidly through a full ROM (ACSM 2021; Gibson, Wagner, and Heyward 2019; Merrill 2015). Some modifiable and nonmodifiable factors that can affect flexibility are age, sex, physical activity and fitness level, chronic disease and other conditions (e.g., stroke, cerebral palsy, etc.), previous injury, and restrictions due to joint structure (Gibson, Wagner, and Heyward 2019; Merrill 2015). To increase the ROM of a joint, there must be a lengthening of the musculotendinous unit producing viscoelastic changes (Merrill 2015; Sharman, Cresswell, and Riek 2006). This enables the muscle to enhance length when a force, like a stretch, is applied and then return to its original position when that force is removed. This is a property observed in elastic tissues. For permanent length changes in tissues like tendons and fascia (nonelastic), the force applied must be greater than what the tissue is used to experiencing to elicit deformation or a change in the shape of tissue and overall microtrauma (Gibson, Wagner, and Heyward 2019). The tissue will adapt over time, leading to greater extensibility (known as "creep") and less resistance to movement at the joint, improving both ROM and mobility (Merrill 2015). Additionally, length changes in muscle can be produced with flexibility training through inhibition of muscle spindles and desensitization of the stretch reflex, which is modulated by the muscle spindles. This leads to a decrease in tension and stiffness within the musculotendinous unit called stress relaxation (Gibson, Wagner, and Heyward 2019; Merrill 2015). There is also evidence to suggest that stretch tolerance (pain associated with tension during a muscle stretch) increases with training, allowing the muscle to lengthen further before the individual experiences pain. These changes require consistent training and healthy stress to tissues through a comprehensive flexibility program.

The primary purpose of this lab is to introduce you to three common methods involved in the assessment of flexibility and ROM: modified sit-and-reach test, double inclinometer test, and goniometer tests. Also, you will be provided with opportunities to perform and administer these assessments in a lab. During these experiences, you will also practice palpation techniques to find bony landmarks; thus, please review anatomical information from courses that you may have taken previously if needed. The secondary purpose of this lab is to teach the principles behind the design of flexibility and ROM programs using the frequency, intensity, time, and type (FITT) method recommended by ACSM (2021). A case study will be given to provide you with an opportunity to apply these principles and interpret flexibility and ROM data.

Necessary Equipment or Materials

- Large goniometer
- Sit-and-reach box with yard or meter stick
- Inclinometer × 2
- Nonpermanent marker to mark double inclinometer sites
- Cushioned floor mat and table (therapy table preferred)
- Flexibility and range of motion data sheet
- Flexibility and range of motion exercise program planning sheet

Calibration of Equipment

For this lab, no calibration of equipment is needed. Make sure that all equipment is in good working condition before use.

Procedures

Because flexibility is very specific to the joint, its structure, and the tissues around it, no single test exists that reflects whole body flexibility (Gibson, Wagner, and Heyward 2019). It is recommended to determine flexibility and ROM through a series of tests involving multiple joints. Thus, in this lab you will be assessing static flexibility using both direct measures (inclinometer test and goniometry) and indirect measures (modified sit-and-reach) at various sites around the body. After administering these tests, the exercise professional should have a reasonable picture of their client's overall flexibility.

The main setup for the lab requires students to work in small groups (2-4 students) to collect data for each flexibility and range of motion assessment. You can complete the assessments in the order of your choosing. Please record your own individual data on the data collection sheet and compare the results of each assessment to the normative data tables. Think about an interpretation of the results and identify areas where you need to improve upon or maintain flexibility and ROM.

Following assessment and data analysis, you will work on the sample case study and design a basic flexibility program. Use the exercise program planning sheet to fill in the information that you choose to prescribe for the client based on their goals and demographical information.

Modified Sit-and-Reach Test

The modified sit-and-reach test was designed to account for relative differences in upper and lower body limb length among individuals that the standard sit-and-reach test does not address (Hoeger and Hopkins 1992). The test serves as an indirect measure of static flexibility of the lower back and hamstring group (Gibson, Wagner, and Heyward 2019). However, research evidence shows that the sit-and-reach test is a modest indicator of hamstring extensibility (Lemmink et al. 2003; López-Miñarro, Andújar, and Rodríguez-García 2009) and lower back flexibility (Lemmink et al. 2003; Martin et al. 1998). Plus, it is a poor predictor of current and future risk of low back pain (Grenier, Russell, and McGill 2003; Jackson et al. 1998). Although past research calls into question the overall validity of the sit-and-reach test, it remains a useful tool for evaluating hamstring flexibility and identifying individuals who may have either very poor flexibility or possess very hypermobile joints (Gibson, Wagner, and Heyward 2019). The procedures for this test are as follows:

1. Instruct your client to perform a general, dynamic warm-up (i.e., walk or slow jog) of the muscles that are involved in the flexibility assessment followed by a light static stretch of each muscle group.

2. Place the sit-and-reach box on a floor mat and make sure that the area is flat. The client should remove their shoes and put their feet flat against the sit-and-reach box. The back of their knees should be in contact with the mat and their low back area, shoulders, and head should be in contact with the wall.

3. Place a large yard or meter stick on top of the sit-and-reach box with the zero end nearest the client. While keeping their low back, shoulders, and head in contact with the wall, instruct the client to raise their arms straight in front of their body with their hands on top of one another. Position the measuring stick so that it is in contact with their middle fingers. This is the starting point for measurement.

4. Secure the measuring stick to the top of the sit-and-reach box and instruct your client to slowly reach forward while bending at the waist and keeping their legs flat against the mat. Both hands should remain evenly stacked together. The client should exhale and drop their head between their arms as they move their hands forward. Once they

reach the end point, they will hold this position for 2 s and then relax.

5. The test should be performed three times in total and each score recorded on the data sheet. Scores should be measured to the nearest 0.25 in. (0.6 cm).

Double Inclinometer Test

The double inclinometer test was designed to be a direct measure of static flexibility of the spine, specifically in the lumbosacral region (Gibson, Wagner, and Heyward 2019). Research evidence supports the use of inclinometer devices to measure ROM in the lumbar spine (Kolber et al. 2013; Mayer et al. 1984; Ng et al. 2001; Saur et al. 1996). Interrater reliability for this measure as an indicator of isolated lumbar flexion and extension is high according to Kolber et al. (2013), Ng et al. (2001), and Saur et al. (1996). Additionally, criterion validity for this measure as an indicator of lumbar ROM is good (Mayer et al. 1984; Saur et al. 1996).

The procedures for this test are as follows:

1. Instruct your client to perform a general, dynamic warm-up (i.e., walk) followed by a repeated series of slow flexion and extension movements of the spine (i.e., cat and camel stretch) with both hands and knees on a floor mat. They should do 10 cycles of each flexion and extension movement.

2. Following the warm-up, the client stands in an upright and comfortable position with their arms to their sides or to their front (figure 9.1a). From this position, you will identify and mark the T12 (twelfth thoracic vertebra) spinous process and the S1 spinous process (just below fifth lumbar vertebra or L5).

3. To find T12, locate the last (twelfth) rib, follow it to the spine, and palpate the process at this spot. Mark the spot with a marker. To find S1, palpate the iliac crest on the back of the client with both hands and touch the tip of your thumbs together. This is the position for L4-L5. Move down the spine over the L5 spinous process to the next position. This is S1. Mark this spot with a marker.

4. Center and place each inclinometer over the marked spots. Zero the inclinometer

out by spinning the dial to where the zero line matches with the top of the "bubble" or fluid in the inclinometer.

5. Instruct the client to bend at the waist as far as they can while reaching to the floor in a slow, smooth motion. Their knees should remain in extension during the movement. Once to their limit, the client should hold this position momentarily until you obtain a reading on both inclinometers (figure 9.1*b*). It is helpful to have the assistance of another technician to read the inclinometers because holding the devices in place and reading them at the same time can be difficult. Measure the value to the nearest 1 degree.

6. Once the data are collected, the client can return to the start position and repeat when ready. The movement is performed a total of three times.

7. The three different measurements obtained are total flexion (top inclinometer), pelvic motion (bottom inclinometer), and true or isolated lumbar flexion (the difference between the two inclinometer values). Record each trial on the data collection sheet and identify the best value for each measurement. For this lab, you will only analyze the true lumbar flexion number. A normal lumbar flexion range is between 40 and 60 degrees; thus, you will be either within the range, below, or above.

Goniometer Tests

Goniometry Measurement: Hip Flexion

Goniometry Measurement: Knee Flexion

Goniometry Measurement: Shoulder Flexion

Goniometry Measurement: Shoulder Abduction

Goniometry is a common method used to evaluate joint angles and is an excellent direct measure of static flexibility and ROM specific to the joint being measured (Shultz, Houglum, and Perrin 2015). The ACSM (2021) recommends the use of goniometer devices to best measure joint ROM. A goniometer resembles a protractor and includes two arms—a stationary or stabilization arm and a movement or measurement arm—and an axis scale that contains degree values that are read during measurement. In this lab, joint ROM will be recorded as the difference between the angle from the starting position to the angle at the end range position during passive motion (Gibson, Wagner, and Heyward 2019).

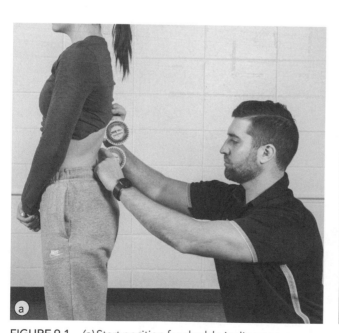 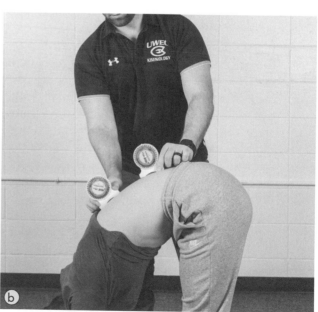

FIGURE 9.1   *(a)* Start position for double inclinometer measurement; *(b)* end position for double inclinometer measurement.

The procedures for this test are as follows:

1. Instruct your client to perform a general, dynamic warm-up (i.e., walk or slow jog for lower body and arm circles for upper body) of the muscles that are involved in the flexibility assessment followed by a light static stretch of each muscle group.

2. Review all goniometry procedures in table 9.1 prior to beginning the tests. Rotate through the goniometry measurements in the order of your choosing. Make sure that your client is in a comfortable position for all measurements.

3. In general, position both arms along the longitudinal axis of either the stabilized body segment (stabilization arm) or the moving body segment (movement arm) and align the axis of rotation on the goniometer with

TABLE 9.1   Standard Procedures for Goniometry Measurement in Select Joints

Joint and movement direction	Body position	Axis of rotation	Stabilization arm	Movement arm	Body part(s) to stabilize	Other notes
Ankle: dorsiflexion and plantar flexion	Seated	Lateral malleolus	Head of fibula is reference point	Align parallel with fifth metatarsal	Lower leg	Legs hang free from table with knee flexed and ankle at 90 degrees
Knee: flexion	Supine	Lateral epicondyle of femur	Midline of femur; greater trochanter is reference point	Midline of fibula; align with lateral malleolus	Femur (prevent rotation, abduction, and adduction)	Allow hip to flex
Hip: flexion and extension	Supine (flex) and prone (ext)	Greater trochanter	Midline of trunk	Midline of femur, lateral epicondyle is reference point	Pelvis	Knee flexes with hip flexion; knee is flexed with hip extension
Hip: abduction and adduction	Supine	Anterior superior iliac spine (ASIS)	Align with opposite side of ASIS	Midline of femur, align with center of patella	Pelvis	Extend knee during abduction; abduct opposite hip to clear room for adduction movement
Hip: internal rotation and external rotation	Seated	Center of patella	Aligned vertically	Anterior midline of lower leg, crest of tibia is reference point	Distal end of femur (no rotation or lateral tilt of pelvis)	Seated on table for legs to hang, both knees at 90 degrees flexion
Shoulder: flexion	Supine	Center of humeral head near acromion process	Midline of thorax	Midline of humerus (lateral epicondyle alignment)	Scapula	Elbow is slightly flexed, palm up; no adduction, abduction, or rotation of shoulder

> continued

TABLE 9.1   Standard Procedures for Goniometry Measurement in Select Joints   > *continued*

Joint and movement direction	Body position	Axis of rotation	Stabilization arm	Movement arm	Body part(s) to stabilize	Other notes
Shoulder: extension	Prone	Center of humeral head near acromion process	Parallel to midaxillary line	Midline of humerus (lateral epicondyle alignment)	Scapula	Palm faces inward; no adduction, abduction, or rotation of shoulder
Shoulder: abduction	Supine	Center of humeral head near acromion process	Parallel to sternum	Midline of humerus	Scapula and thorax	Palm faces anteriorly; humerus laterally rotated; elbow extended; no flexion or extension of shoulder
Shoulder: internal rotation and external rotation	Supine	Olecranon process	Vertical alignment	Styloid process of ulna	Distal end of humerus and scapula	Shoulder 90 degrees abduction, elbow flexed 90 degrees, forearm is perpendicular to floor and in neutral position; rest humerus on mat and keep level with acromion process

Adapted by permission from A.L. Gibson, D.R. Wagner, and V.H. Heyward, *Advanced Fitness Assessment and Exercise Prescription*, 8th ed. (Champaign, IL: Human Kinetics, 2019), pp. 315-316.

each joint's axis of rotation (Gibson, Wagner, and Heyward 2019; Shultz, Houglum, and Perrin 2015).

4. Make sure to position yourself so that you are close to eye level with the goniometer scale. Before reading the scale, check to make sure that both arms and the axis of rotation of the goniometer are in proper alignment (Shultz, Houglum, and Perrin 2015).

5. Since passive ROM is being measured, move the client's limb through the full ROM starting at the zero position and going to the end point of the ROM. Ensure that the proximal portion of the limb is stabilized before beginning movement (Shultz, Houglum, and Perrin 2015).

6. Record the degrees on the goniometer scale at the end of the ROM and repeat the movement for a total of three trials.

Common Errors and Assumptions

As we have seen, the measurement of flexibility and ROM can be accomplished in a few different ways. Some are very simple to use and take very little skill while others take more practice and time to learn. The choice of modality may be influenced by issues such as skill of the exercise professional, accessibility of the modality itself, and the specific movement that we may want to evaluate (Kolber et al. 2013). Whenever a greater amount of skill to perform a test is involved, the chance for error increases. As always, following the specific procedures for each modality and improving skill proficiency through practice is important to reduce error and increase validity and reliability. The following sections address sources of measurement error for all flexibility and ROM methods.

Common Errors in Modified Sit-and-Reach Test

A major advantage is that this test requires very little technical skill on the part of the exercise professional. Therefore, most of the error lies with test technique and performance of the client. Two common errors are poor posture (i.e., head, back, and hips not in contact with the wall) of the client at the starting point and the legs not remaining in contact with the floor during forward flexion. Another common technique error occurs when one hand begins to lead in front of the other during measurement. This will elicit a greater flexibility value compared to the actual value. To eliminate these errors, be very clear in your test directions, cue technique appropriately, and remain observant to correct errors as they occur.

Common Errors in Double Inclinometer Test

Some advantages for using inclinometers to measure spine flexion are their portability, low cost, and lower skill requirement compared to goniometry (Kolber et al. 2013). However, the exercise professional's technical proficiency is the main source of error during this measurement (Samo et al. 1997). The most common error is poor identification of anatomical landmarks and subsequent misplacement of the inclinometer. Additionally, it is important to maintain position of the inclinometers to reduce measurement error during movement; therefore, hold them firmly and vertically along the line of the spine (Kolber et al.

2013). To reduce these errors, learn palpation techniques from a skilled technician and practice this method on a variety of individuals with different body types and ages.

Common Errors in Goniometer Tests

The major advantage of goniometry is that it involves a single, inexpensive tool to precisely measure joint ROM (Kolber et al. 2013). Just like with inclinometry, the main source of error in goniometry is the technical skill of the exercise professional. Common errors that affect the validity of results include improper positioning of the client, goniometer misalignment based on anatomical sites, improper stabilization of the proximal limb or segment during movement, not moving the client through full ROM, and not reading the goniometer at eye level (Shultz, Houglum, and Perrin 2015). To reduce these errors, learn goniometry techniques from a skilled technician and practice this method on a variety of individuals with different body types and ages.

Normative Data and Interpretation

Tables 9.2 and 9.3 contain the normative flexibility values specific to men and women of varying age groups and the average ROM values, respectively. You will use these normative charts to classify your flexibility and ROM data. Table 9.2 will be used to analyze the results of the modified sit-and-reach

TABLE 9.2   Classification of Results for the Modified Sit-and-Reach Test

Age (yr)	Well above average		Above average		Average		Below average		Well below average	
MEN	IN.	CM	IN.	CM	IN.	CM	IN.	CM	IN.	CM
≤35	≥17.2	≥43.7	15.8-17.1	40.1-43.4	13-15.7	33-39.9	9.2-12.9	23.4-32.8	≤9.1	≤23.1
36-49	≥16.1	≥40.9	13.9-16	35.3-40.6	10.8-13.8	27.4-35.1	8.3-10.7	21.1-27.2	≤8.2	≤20.8
≥50	≥15	≥38.1	12.3-14.9	31.2-37.8	9.3-12.2	23.6-31	7.8-9.2	19.8-23.4	≤7.7	≤19.6
WOMEN	IN.	CM	IN.	CM	IN.	CM	IN.	CM	IN.	CM
≤35	≥17.9	≥45.5	16.2-17.8	41.1-45.2	13.7-16.1	34.8-40.9	10.1-13.6	25.7-34.5	≤10	≤25.4
36-49	>17.4	≥44.2	15.2-17.3	38.6-43.9	12.2-15.1	31-38.4	9.7-12.1	24.6-30.7	≤9.6	≤24.4
≥50	≥15	≥38.1	13.6-14.9	34.5-37.8	9.2-13.5	23.4-34.3	7.5-9.1	19.1-23.1	≤7.4	≤18.8

Adapted by permission from A.L. Gibson, D.R. Wagner, and V.H. Heyward, *Advanced Fitness Assessment and Exercise Prescription*, 8th ed. (Champaign, IL: Human Kinetics, 2019), 323. Based on Hoeger and Hoeger (2015) and Beam and Adams (2011).

TABLE 9.3   Ranges of Average ROM Values From Goniometry

Joint and movement direction	ROM (deg)	Joint and movement direction	ROM (deg)
Ankle: dorsiflexion	15-20	Hip: internal rotation	40-45
Ankle: plantar flexion	45-50	Hip: external rotation	45-50
Knee: flexion	135-150	Shoulder: flexion	150-180
Hip: flexion	100-120	Shoulder: extension	50-60
Hip: extension	25-30	Shoulder: abduction	170-180
Hip: abduction	40-45	Shoulder: internal rotation	60-70
Hip: adduction	20-30	Shoulder: external rotation	80-90

Note: All goniometry measurements start at the zero point and are measured in degrees.

Data from Greene and Heckman (1994) and Kendall et al. (2005).

test and table 9.3 for the results of the goniometry measurements. For table 9.2, find your sex-specific section and age-specific row to determine your flexibility classification using the terms at the top of the chart. For table 9.3, compare your goniometry measurements to the average ranges and then determine if you fall within range, below (less ROM), or above (more ROM). Take these findings, along with your inclinometer test results, and develop an overall picture of your flexibility and ROM. Where are your trouble spots that should get attention? Which joints fall within normal limits of ROM or which are hypermobile? It is very helpful to gather flexibility and ROM data from various parts of the body to get a full clinical picture of joint mobility. Again, one single test does not give us all the answers.

While it is appropriate to evaluate flexibility and ROM data compared to those of similar age and sex, these data should also be interpreted with a functional fitness and joint mobility focus in mind. As discussed earlier in this lab, maintaining flexibility and joint mobility across the lifespan is associated with a lower risk of chronic pain, disability, and dependency (Gibson, Wagner, and Heyward 2019; Jones and Knapik 1999). This is especially important for older adults, who should be encouraged to maintain joint mobility for independent living, performing activities of daily living, and overall physical function at home and in the community. Thus, any discussion with a client should include these talking points: (1) where they are currently at from a flexibility and joint mobility standpoint compared to others (classification), (2) what these data mean for their overall physical function and mobility (interpretation), (3) whether they should either maintain or improve their current level of flexibility and at what joints or areas

(interpretation), and (4) what they should do to achieve their goals (interpretation and planning of exercise program). By focusing on these points, you will be able to make the biggest impact on your client's understanding of the data and help them plan an appropriate exercise program.

Exercise Program Design: Flexibility and ROM

The main purpose of the flexibility and ROM prescription is to provide a gradual exercise stimulus over time to deform connective tissue and provide a neuroinhibitory effect on muscle to improve joint mobility and flexibility in muscle and tendon groups (ACSM 2021). Other individual goals in a flexibility exercise program could include decreasing chronic pain, disability, or stress, and increasing muscle relaxation, improving posture and balance, improving movement efficiency and physical function in aging individuals and those with disabilities, reducing joint restrictions and contractures, or improving sport performance (ACSM 2021; Behm et al. 2016; Garber et al. 2011; Glei et al. 2019; Gordon et al. 2004; Kokkonen et al. 2007; Merrill 2015). Remember, no matter what the goals are, the exercise professional should always encourage clients to set their own goals, assist them to further clarify and define goals, and design an individualized exercise program around these goals. Individualization is key for program adherence and regular participation in exercise.

This section will provide you with information to design a basic flexibility program for clients seeking to improve their overall flexibility and ROM. Before any type of flexibility training (except for dynamic flexibility), a generalized warm-up to increase blood flow and temperature within the muscle is encouraged to increase the effectiveness of the flexibility exercise (ACSM 2021; Garber et al. 2011).

Frequency

Frequency of exercise refers to the number of days a client engages in flexibility training per week. ACSM (2021) recommends a training frequency of at least 2 to 3 d·wk^{-1}, with daily flexibility exercise being the most effective at improving ROM. ROM of a joint can be acutely improved following an exercise session and chronically improved after a month of flexibility exercise through as little as two to three times per week (Garber et al. 2011).

Intensity

Intensity refers to the tension that is felt in the muscle during the act of stretching. ACSM (2021) recommends moving through a stretch to the point of sensing tightness or slight discomfort in the muscle. Once the individual reaches this point, they should hold for a designated period of time. The overall time that this position is held varies among different types or modes of flexibility training.

Time and Volume

In this discussion, time refers to the duration (usually reported in seconds) each stretch is held at the point of tension or mild discomfort in the muscle. Volume refers to the total time of stretching that is completed per joint. For static stretching, each stretch should be held for 10 to 30 s for adults and longer (30-60 s) for older individuals to elicit improvements in ROM (ACSM 2021; Garber et al. 2011). For proprioceptive neuromuscular facilitation (PNF) exercises, time of hold during a stretch is also 10 to 30 s in length (ACSM 2021). Per joint, the total volume of time with the muscle under stretch should be 90 s. Other modes of flexibility training differ in time and will be addressed in their respective sections.

Type (Mode) and Timing of Flexibility Training

Type or mode refers to the specific type of flexibility exercise that is performed. It is recommended to use a variety of flexibility techniques to improve joint ROM based on their individual effectiveness for the client, the skill and knowledge that the client has for each mode, and availability of either a partner or equipment to perform the specific mode. In this section, timing refers to when the flexibility training is performed. For example, flexibility exercise could be performed prior to exercise such as aerobic or strength training, an athletic event, or an occupational activity. Additionally, flexibility training could be done either at some point following these activities or as a stand-alone program. Types of flexibility exercise and recommendations for timing are as follows.

Static Stretching

The act of static stretching involves gradual stretching of a muscle to a point of tension that is held for a specified time. This action is repeated as often as needed to meet the volume goal for the client. Advantages of static stretching are that it can be done alone, it is simple and less strenuous for the client to do, and it is less likely that the client will go beyond normal end ranges of the specific motion (Merrill 2015). Static stretching can be either active (holding a stretched position using other muscles) or passive (holding a stretched position with or without assistance) (ACSM 2021; Garber et al. 2011).

Static stretching has been shown to be an effective modality to increase joint ROM due to modulation of musculotendinous unit stiffness, neural inhibition, and improved stretch tolerance (ACSM 2021; Behm et al. 2016; Borges et al. 2018; Garber et al. 2011; Kokkonen et al. 2007). Because of these changes, the timing of static stretching is important for certain types of exercise performance, namely strength- and power-related activities (Behm et al. 2016; Behm and Chaouachi 2011; Garber et al. 2011; Kay and Blazevich 2012; Shrier 2004). In general, it is recommended to avoid static stretching immediately prior to activities that involve strength and power due to its negative effects on exercise performance. Thus, static stretch training to increase ROM and/or improve performance, as shown by Kokkonen and colleagues (2007), is best planned as a separate program (Behm et al. 2016; Garber et al. 2011). However, if static stretching is performed prior to these activities, the duration and volume of stretching should be ≤60 s in length to minimize any detrimental effects (Behm et al. 2016; Kay and Blazevich 2012).

Proprioceptive Neuromuscular Facilitation

PNF stretching involves repeated cycles of passive stretching to a point of tension and active contraction of the muscle. This action is repeated as often as needed to meet the set volume goal for the client. The main advantages of PNF stretching are that it appears to be an effective method at improving joint ROM, both passive and active, and can be performed in multiple planes (Lempke et al. 2018; Merrill 2015; Sharman, Cresswell, and Riek 2006).

PNF stretching works to increase joint ROM through lengthening of the musculotendinous unit and neural inhibition (Konrad, Stafilidis, and Tilp 2017; Sharman, Cresswell, and Riek 2006). This can be done through two means of PNF stretching: contract-relax (CR) and contract-relax-agonist-contract (CRAC) (Merrill 2015; Sharman, Cresswell, and Riek 2006). To perform a PNF stretch using CR (autogenic inhibition), the exercise professional slowly stretches the targeted muscle into a position of mild discomfort for 10 to 30 s. Following this, the client is instructed to contract the targeted muscle for 3 to 6 s and then relax followed by another 10 to 30 s of deeper passive stretching of the targeted muscle. For PNF stretching using CRAC (reciprocal inhibition), the exercise professional slowly stretches the targeted muscle into a position of mild discomfort and tension for 10 to 30 s. Afterward, the client is instructed to contract the opposite muscle (i.e., stretch the hamstrings and contract the quadriceps) for 3 to 6 s, followed by another 3 to 6 s of contraction of the targeted muscle and then 10 to 30 s of deeper passive stretching of the targeted muscle. An intensity of contraction of 20% to 75% (light to moderate) of maximal voluntary contraction has been recommended in the research literature, although as little as 20% of maximal voluntary contraction has been deemed sufficient to increase joint ROM (ACSM 2021; Garber et al. 2011; Sharman, Cresswell, and Riek 2006).

Both methods of PNF stretching produce improvements in joint ROM, with CRAC thought to be the more effective method (Sharman, Cresswell, and Riek 2006). Also, because of the changes in the muscle induced by PNF exercise, the timing of this modality with other activities is similar to static stretching (ACSM 2021; Behm et al. 2016). Therefore, the planning of PNF training is best either after exercise and other activities or as a separate program altogether.

Active Isolated Stretching

Active isolated stretching (AIS) is a stretching technique that combines aspects of both active and passive stretching (Merrill 2015). To begin, a client actively moves the target muscle to a point of mild discomfort and holds this position for ≤2 s. Once that time is up, the muscle is relaxed and returned to the starting position. This cycle is repeated 5 to 10 times with the client going deeper into every subsequent stretch. The stretch can be done through actively moving the limb or with assistance using a Theraband or similar tool. Repetition number is dependent upon the client and their overall goals. While not much work has been done on this technique, an advantage could be its short duration of position hold, which may make it less likely to cause any negative effects on exercise performance if done prior to activity.

Dynamic Stretching

Dynamic stretching, also referred to as dynamic range of motion (DROM), utilizes controlled movements that mimic a particular functional skill or sport-specific movement in order to improve ROM while at the same time enhancing joint stability, coordination, and muscle activation (ACSM 2021; Behm et al. 2016; Behm and Chaouachi 2011; Garber et al. 2011; Merrill 2015). Dynamic stretching has been found to be an effective method for preparing the body for movement during activity; thus, it is a recommended part of any pre-exercise warm-up routine (ACSM 2021; Behm et al. 2016; Behm and Chaouachi 2011; Garber et al. 2011). If this method is new to your client, prescribe very simple exercises that do not require a high level of skill and intensity to perform. Progress to more complex, multiplanar movements at higher intensity levels as the client gains more experience and fitness (Merrill 2015). DROM can also be coupled with static stretching within a warm-up provided that DROM follows static stretching to reduce any potential acute negative effects of static stretching on exercise performance.

Myofascial Release

While myofascial release (MFR) is not a flexibility modality per se, it is a method that can be used to enhance joint ROM. Enhancing flexibility using MFR is thought to occur through external pressure on muscle to relieve tension and reduce restrictions in underlying fascia to improve mobility (Merrill

2015). Muscle tension is released through stimulating mechanoreceptors producing some muscle inhibition leading to relaxation of the muscle and the release of trigger points within the fascia and other soft tissues (Wilke et al. 2020). The benefit for the client is less tension in the muscle and tissues, better posture, and improvement in function and movement. Maladapted fascia is less pliable, associated with a great number of adhesions causing restricted movement, linked to tight muscles, and a potential source of pain and discomfort (Beardsley and Škarabot 2015; Merrill 2015; Schleip and Müller 2013).

The process of MFR includes applying external pressure (usually body weight and gravity) to a tight or tender area, inducing relaxation in the tissue and suppression of pain (Merrill 2015; Wilke et al. 2020). The amount of pressure applied and time it is applied is specific to the client and their overall pain tolerance. In general, clients should move back and forth on a small, tight area (2-6 in. [0.79-2.36 cm] at most) for 30 to 60 s or longer at a time to reach the goal of reduced pain and tension. Once this is achieved, the client should move to a different area of tightness. This is the pattern of MFR training, not rolling continuously up and down the length of a muscle. Implements that are commonly used for MFR are traditional foam rollers of varying density, tennis balls, lacrosse balls, roller sticks, and other tools (Cheatham et al. 2015; Schleip and Müller 2013). Varying the density (soft to hard) can also be effective either for clients with differing degrees of pain tolerance or for use in different areas of the body that are usually more tender than others (i.e., MFR of shoulder muscles versus hamstrings).

Timing of myofascial release has received some attention in the research literature. Generally, most research has shown no adverse changes to exercise performance following MFR training (Beardsley and Škarabot 2015). Thus, MFR can be done at any time that best fits the client in their individual program.

Repetitions per Exercise and Number of Exercises

For static stretching and PNF, it is recommended that clients engage in at least two to four (static) or one to three (PNF) repetitions of a single flexibility exercise (Garber et al. 2011; Sharman, Cresswell, and Riek 2006). Remember, the goal is to get at least 90 s of stretching volume per exercise. The time of each repetition should be planned with this volume goal in mind. The number of exercises recommended is at least one exercise per major joint (Garber et al. 2011).

For MFR, the number of repetitions is dictated by a noticeable change in point or general tenderness and muscle tension (Merrill 2015). Thus, the number of repetitions with this modality is client dependent and continued until the goal is achieved. For DROM, one set of up to 10 to 20 repetitions per exercise is generally recommended. The number of exercises chosen is dependent upon what areas of the body need to be prepared for the subsequent activity. For higher tempo and intensity DROM exercises, repetition ranges may need to be adjusted down. Finally, for AIS, one to two sets of up to 10 repetitions per joint is recommended.

Case Study for Flexibility and ROM Exercise Prescription

For the program design, you will fill in the flexibility and range of motion exercise program planning sheet with your recommendations for type of exercise, number of sets and repetitions performed for each exercise, time of hold for the stretch or time the movement is performed (depending on the exercise), and rest periods between repetitions. For type of flexibility and ROM exercises, focus on DROM exercises for the pre-exercise period (some examples have been provided) and static, PNF, AIS, and MFR exercises for the post-exercise period. Remember, timing of the program can be immediately after or at a different time of day depending on the client's needs. For simplicity, this prescription is for a client who is new to flexibility exercise and at the initial stage of their program. You can choose up to 15 different exercises from the different types of flexibility and ROM training modes. For examples of select flexibility exercises, review appendix F.1 in the *Advanced Fitness Assessment and Exercise Prescription* text by Gibson, Wagner, and Heyward (2019).

As in lab 7, with a real client, you would have the opportunity for a back-and-forth conversation and communication as the program is developed and progressed over time. However, do your best to apply the flexibility program design principles through this case study example and hand in your work to get feedback on your overall plan.

Demography

Age: 50
Height: 5 ft 11 in. (180.3 cm)
Weight: 205 lb (93.2 kg)
Sex: Male
Race or Ethnicity: Hispanic or Latino

Family History

Your client's family history reveals no cardiovascular disease. However, there is a history of breast and lung cancer on the mother's side (mother, sister, and grandmother all before age 65) and their 75-year-old father was diagnosed with type 2 diabetes at age 49.

Medical History

Present Conditions

On the health history questionnaire form, your client reports transient mild to moderate pain and discomfort in several joints and muscles in the upper and lower body (shoulders, low back, hips, knees, and ankles). This occurs at work and during various daily activities around the house. Your client takes ibuprofen and Aleve to manage the pain. A recent physical assessment revealed the following health information. Your client's resting HR was 63 beats·min^{-1} and their resting BP was 126/82 mmHg (confirmed from a previous measurement). The client's skinfold assessment showed a percent body fat (%BF) of 27%. Fasting blood glucose is 85 mg·dL^{-1}. No blood chemistry panel for cholesterol is reported at this time. The most current exercise test, which was a submaximal walking test, showed a $\dot{V}O_2$max value of 26 mL·kg^{-1}·min^{-1}. The flexibility and ROM information you collected follows. No other data are available currently.

Modified sit-and-reach test	Score (cm)	Rating	Double inclinometer test	Score (deg)	Rating
Static flexibility	16		True lumbar flexion	35	

Goniometer test	Score (deg)	Rating	Goniometer test	Score (deg)	Rating
Dorsiflexion	12		Hip internal rotation	32	
Plantar flexion	35		Hip external rotation	44	
Knee flexion	100		Shoulder flexion	130	
Hip flexion	95		Shoulder extension	35	
Hip extension	15		Shoulder abduction	175	
Hip abduction	35		Shoulder internal rotation	55	
Hip adduction	10		Shoulder external rotation	80	

Past Conditions

Your client reports two surgeries, one on the knee (15 yr ago) and one on the ankle (20 yr ago), due to past injuries from sport participation. Otherwise, your client reports no other medical and health issues.

Behavior and Risk Assessment

Your client is a former smoker (they quit 3 yr ago) who has not done regular physical activity for the past 10 yr. Occupationally, they have a very sedentary office job that includes long periods of sitting that cause flare-ups of pain and discomfort in the joints. Your client is coming to you to develop a structured flexibility exercise program to accompany the cardiorespiratory fitness and strength program that you have already made for them. The program that you have designed includes 3 to 4 d·wk^{-1} of moderate intensity physical activity, mostly walking on a treadmill or outside for 35 to 45 min·d^{-1}, and 2 d·wk^{-1} of moderate intensity resistance training at the company fitness facility. This will increase to 5 d·wk^{-1} of walking by the end of the program. The main goals for the flexibility program are to decrease pain and discomfort from what they think are "tight muscles and joints," be above average on all flexibility and ROM assessments by the end of the 12 wk program, and use these exercises to relax and de-stress at night. Your client believes that they could fit flexibility and ROM training in anywhere from 3 to 5 d·wk^{-1} and is open to your suggestions on the exact number of days. There is no information collected on your client's diet.

Review your client's data and provide a rating for each of the assessments in the data table. Also, please fill out the program planning sheet with exercises and other information for your client to follow based on their exercise goals.

FLEXIBILITY AND RANGE OF MOTION DATA SHEET

Name _____ Sex _____ Date _____

Age _____   Height _____ cm   Weight _____ kg

Modified Sit-and-Reach Test

Trial 1 ____ cm ____ in.   Trial 2 ____ cm ____ in.   Trial 3 ____ cm ____ in.

Note: Take the best of three trials and analyze. Values should be recorded in cm and in.

Percentile rank and rating from table 9.2 _____

Double Inclinometer Test

Total flexion (top inclinometer)		
Trial 1 _____ deg	Trial 2 _____ deg	Trial 3 _____ deg
Pelvic (sacral) motion (bottom inclinometer)		
Trial 1 _____ deg	Trial 2 _____ deg	Trial 3 _____ deg
True (isolated) lumbar flexion (difference between both measurements)		
Trial 1 _____ deg	Trial 2 _____ deg	Trial 3 _____ deg

Note: Take the best of three trials and analyze. Values are recorded in degrees.

Rating for true lumbar flexion _____

Goniometer ROM Tests

Goniometry measurement	Trial 1	Trial 2	Trial 3	Rating
Dorsiflexion				
Plantar flexion				
Knee flexion				
Hip flexion				
Hip extension				
Hip abduction				
Hip adduction				
Hip internal rotation				
Hip external rotation				
Shoulder flexion				
Shoulder extension				
Shoulder abduction				
Shoulder internal rotation				
Shoulder external rotation				

Note: Circle the best of three trials and analyze based on table 9.3. Values are recorded in degrees.

FLEXIBILITY AND RANGE OF MOTION EXERCISE PROGRAM PLANNING SHEET

Dynamic Flexibility Exercises (Pre-Exercise)

Exercise	Sets	Repetitions	Time (s)	Rest (s)
Lunge with overhead side reach	2	20 (10 per leg)	10	15
Knee to chest (walking)	2	20 (10 per leg)	5	15
Side lunges	2	20 (10 per leg)	5	15
Inchworm	2	10	NA	15
Forward lunge with elbow to instep	2	20 (10 per leg)	5	15

Flexibility and ROM Exercises (Post-Exercise or Other Time)

Frequency _____ d·wk^{-1}

Exercise	Sets	Repetitions	Time (s)	Rest (s)

From J. Janot and N. Beltz, *Laboratory Assessment and Exercise Prescription* (Champaign, IL: Human Kinetics, 2023).

Assessment of Muscular Fitness: Muscular Endurance

Purpose of the Lab

Muscular endurance is the ability of a muscle or muscle group to execute repeated contractions against resistance that leads to muscular fatigue or failure (Gibson, Wagner, and Heyward 2019; Ratamess et al. 2009). Tests used to measure muscular endurance are often dynamic in nature, such as the curl-up test, the push-up test, and the pull-up test, among others. Muscular endurance can also be evaluated by the total time required to either hold a static posture or a percentage of maximum voluntary contraction (MVC) until fatigue or failure. A key piece to all effective endurance tests is that they are open-ended, meaning that there is no time limit for completion. For instance, a 1 min push-up test could be a good test for an individual who fatigues before the time limit is up. But if that individual could continue beyond that minute, the result would underestimate their

endurance potential. Thus, the goal is to complete as many repetitions as possible or hold a position until an end point of muscular failure, and not to achieve a specific number of repetitions or reach a specific time limit. This is all completed with no rest periods or breaks during the test.

The main portion of this lab will be spent on evaluating core endurance. Appropriate levels of muscular endurance have been linked to better performance of activities of daily living, improved dynamic stability and postural control, enhanced sport performance, and decreased risk of back or spine issues, among other benefits (Gibson, Wagner, and Heyward 2019; Magyari 2018; McGill 2015; Ratamess et al. 2009). Regarding back health, it has been demonstrated that the balance of endurance among the anterior, lateral, and posterior spine stabilizing musculature is disrupted once back problems occur (McGill et al. 2003). Often, the endurance of the spine extensors is lower, compared to the muscles of the flexor and lateral groups, in those with back problems who present with recurrent symptoms. Over time, when evaluating the relationships among flexor, extensor, and lateral musculature, the goal is to maintain a balance in endurance performance relative to all areas in order to avoid future risk of back pain. Because of this, it is good practice to administer more than one test, because a single test will not provide enough information due to the testing specificity of muscular endurance (McGill 2015). Therefore, we will need a whole battery of tests to give us a clearer picture of whole body muscular endurance.

The primary purpose of this lab is to introduce you to methods involved in the assessment of muscular endurance: the YMCA bench press test and the core endurance test battery. Also, you will be provided with opportunities to perform and administer these assessments in the lab and will be required to apply spotting techniques to keep your client safe. Some of these techniques will be used later in lab 11 activities. The secondary purpose of this lab is to teach the principles behind the design of core endurance programs using the FITT method recommended by ACSM (2021). A case study will be given as an opportunity for you to apply these principles and interpret muscular endurance data.

Necessary Equipment or Materials

- Cushioned floor mat and table (therapy table preferred)
- Yard or meter stick and marking tape
- Stopwatch and metronome
- Flat bench and barbell (up to 80 lb [36.4 kg] of weight)
- Flat wooden board or plastic bench step
- Muscular endurance data sheet
- Core endurance exercise program planning sheet

Calibration of Equipment

For this lab, no calibration of equipment is needed. Make sure that all equipment is in good working condition before use.

Procedures

Because muscular endurance is very specific to the muscle or muscle groups involved, no single test can represent overall muscular endurance as a component of fitness (Gibson, Wagner, and Heyward 2019; McGill 2015). It is recommended to determine muscular endurance through multiple assessments. Thus, in this lab, you will be assessing both upper body (YMCA bench press test) and core (core endurance test battery) endurance. After administering these tests, the exercise professional should have a reasonable picture of their client's overall muscular endurance.

The main setup for the lab requires students to work in small groups (2-4 students) to collect data for each muscular endurance assessment. You can complete the assessments in the order of your choosing. To prevent carryover fatigue from test to test, give clients at least 5 min of rest between each test. Please record your own individual data on the data collection sheet and compare the results of each assessment to the normative data. Think about an interpretation of the results and identify areas that either need to be improved or maintained regarding muscular endurance.

Following assessment and data analysis, you will work on the sample case study and design a basic core endurance program. Use the exercise program planning sheet to fill in the information that you choose to prescribe for the client based on their goals and demographical information.

YMCA Bench Press Test

The YMCA bench press test was designed as a measure of dynamic upper body muscular endurance (Golding 2000). Free weights (constant load) and a barbell are the assessment tools utilized in this test; thus, clients should be monitored very closely from beginning to end for safety reasons. Spotting techniques for this test include assisting the client with moving the bar off the rack and into the starting position, using an alternating hands grip (i.e., one supinated and one pronated) to control the bar into position, good communication with the client throughout the test (e.g., encourage, motivate, ask how they are doing, etc.), and assisting the client at the end of the test or if a problem arises (Gibson, Wagner, and Heyward 2019). Remember, this is a test with a potential end point of complete muscular failure on the part of the client. The exercise professional should be focused on their client and anticipate intervening if needed. If you are in good communication with your client, the end of a test should never catch you off guard. The procedures for this test are as follows (Golding 2000):

1. Before beginning the test, ask your client to perform a generalized warm-up for 5 min focused on getting the upper body ready to perform the test. This could include light wall push-ups, arm circles, elastic resistance band movements for the arms, and so forth. If your client is unfamiliar with the bench press movement, provide proper instruction and an opportunity for your client to practice (without resistance) before the test.

2. Prepare the bench and barbell setup for your client. The test requires a resistance of 80 lb (36.4 kg) for men and 35 lb (15.9 kg) for women.

3. Set a metronome at 60 beats·min^{-1} to achieve a cadence of 30 repetitions for each minute. This means that the client moves the bar either up or down for each beat of the metronome.

4. The client should start the test in the "up" position and can begin on any metronome beat they choose. The client should perform as many repetitions of the bench press movement as possible. Only "good" (i.e., bar comes down to the chest and back up to the starting position) repetitions done with proper form should be counted. Enlist the help of another group member to count repetitions while you spot and watch form.

5. The test should be stopped once the client can no longer keep up with the cadence of the metronome. No rest is allowed during the test.

Core Endurance Test Battery

In this section, you will learn and practice three core endurance tests—the flexor endurance test, the extensor endurance test, and the lateral endurance test—for both the right and left side of the trunk. The core endurance test battery developed by McGill (2015) is designed to assess the endurance of the core and spine stabilizing musculature in the anterior, lateral, and posterior areas of the

trunk. In the past, the recommended assessments for core endurance were either the bent-knee sit-up or curl-up. However, a main limitation of these tests is that they only assess the endurance of the anterior torso (flexor) musculature, accounting for only one-third of the areas used to stabilize the spine. Clearly, one test does not give us all the information needed to fully evaluate the overall core endurance of these spine stabilizers. Additionally, understanding the balance of endurance among these groups of muscles better informs the exercise professional of current and future risk of back issues with certain clients and identifies areas to target during training (McGill 2015). Therefore, we need to include an assessment for the posterior and lateral trunk musculature to complete the picture.

Because stabilizing the spine during whole body movement includes an isometric action of core muscles, we will isometrically test the endurance of these anterior, lateral, and posterior muscles. These three tests have been shown to have high reliability with repeated measurement over time (McGill, Childs, and Liebenson 1999). Since these tests require effort to near or at failure, screen the client to ensure that the tests will not cause pain if they have a history of acute or chronic low back pain, and avoid them if the client has had recent back surgery or currently has pain. Moreover, avoid the lateral endurance test if your client has current shoulder pain or any past shoulder issues that could be exacerbated by this test.

Before performing the test battery, instruct your client to perform a generalized warm-up of light calisthenics and aerobic activity for 5 to 10 min. The procedures for these tests are as follows (McGill 2015).

Flexor Endurance Test

1. Place a floor mat on a level surface and measure a 10 cm (4 in.) distance between two tape markings on the mat. Also, have a timing device available to you.

2. The client should be seated in a sit-up posture leaning against a board or plastic step (based on availability of equipment) situated at a 55-degree angle from the floor. Align the front edge of the step or board with the first tape mark on the mat. The tester will secure the board or step for the

client throughout the test. Both knees and hips are flexed at 90 degrees and the arms are folded across the chest with each hand on the opposite shoulder. Enlist the help of another group member to hold the client's feet.

3. The test begins when the board or step is pulled back 10 cm (4 in.) to the second tape mark and the person holds the posture for as long as possible. Instruct the client to keep their head aligned with their spine (no forward flexion) and watch for rounding of the shoulders or arching of the back. Also, they may not move their torso forward beyond the starting position. Start timing when the board or step is moved.

4. The test is terminated and time stopped when any part of the client's back touches the board or step. Record the time on the data sheet in seconds.

 ## Extensor Endurance Test

1. For this test, use a cushioned table or place a mat on top of a level table for client comfort and have a timing device available to you.

2. The client will begin with their upper body held out over the end of the table with their hips, knees, and feet in contact with the table. A good reference is making sure that their umbilicus is just over the edge of the table. Enlist the help of a group member to secure the client's legs during the test. Put a chair (no wheels) on the floor at the edge of the table for your client to hold their upper body in a comfortable position before beginning the test.

3. To begin the test, the client will lift their arms and fold them across the chest with each hand on the opposite shoulder. Time is started when the client moves their arms to this position. The client should hold a straight, horizontal posture with their head in alignment through their spine, hips, knees, and feet for as long as possible.

4. The tester should evaluate posture from the side of the client and ensure that the client does not move their torso above the

starting position. As the client fatigues, you may notice a slight drop in their torso below the starting position. Encourage the client to raise their torso back into position if they can.

5. The test is terminated and time stopped when the client can no longer hold the horizontal position. Have the chair in place for the client because they will likely drop their hands and rest them on the chair upon completion of the test. Record the time on the data sheet in seconds.

Lateral Endurance Test

1. Place a floor mat on a level surface and have a timing device available to you.

2. The client should lie on their side with one arm bent 90 degrees at the elbow supporting their upper body off the mat, legs extended out, and feet in a tandem position (i.e., top foot in front of bottom foot). The involved shoulder should be slightly less than 90 degrees abducted to provide better support. The other arm is across the chest and hand placed on the opposite shoulder. Their hip can be in contact with the mat at this time.

3. To begin the test, the client lifts their hips up into a full side bridge position. Time is started when the client moves into this position. The client should hold a straight posture with their head in alignment through their spine, hips, and feet for as long as possible. Only the elbow, forearm, hand, and feet should be in contact with the mat.

4. The tester should evaluate posture from the side of the client and ensure that the client does not move their hips above the starting position. As the client fatigues, you may notice a slight drop in their hips below the starting position. Encourage the client to raise their hips back into position if they can.

5. The test is terminated and time stopped when the client can no longer hold the side bridge position or their hip touches the mat. Record the time on the data sheet in seconds. After sufficient rest, test the opposite side following the same procedures.

Common Errors and Assumptions

As we have seen, the measurement of muscular endurance is accomplished using multiple exercise tests on different areas and muscles in the body. In general, the muscular endurance tests in the lab are relatively easy to administer with simple directions for the client. It is still important to follow the specific procedures for each test, communicate clearly with your client, and practice to improve your proficiency at administering each test, which are the best ways to maintain validity and reliability of these assessments. The following sections address sources of measurement error for the muscular endurance assessments.

Common Errors in the YMCA Bench Press Test

Client performance is the main source of error in the YMCA bench press test. Therefore, instructions and feedback that are given to the client before and during the test need to be clear and accurate. Proper lifting technique is the most important factor for eliciting a good outcome on the test (Gibson, Wagner, and Heyward 2019). Make sure to teach proper bench press technique, give your client an opportunity to practice the bench press motion without resistance, and cue and correct during this practice. The speed of movement will be controlled by a metronome; thus, pay attention to your client's pace and correct if needed. Remember, only repetitions done with good form are counted during the test. For the tester, knowing what weight to use for the client, setting the metronome correctly, and making the correct decision to terminate the test are issues to review prior to beginning the assessment.

Common Errors in the Core Endurance Test Battery

As with the previous test, client performance is the main source of error. This, again, begins with providing clear and accurate instructions before each assessment. The maintenance of appropriate posture during each test is the most important issue to control for to ensure a valid and reliable result. As you review the instructions with your client, put your client through each posture to familiarize them with what will be expected during the test. This will control for any learning effects influencing the outcome of the tests. For the tester, appropriate setup for each test (especially the flexor test), teaching and then cueing proper form during the assessment, and making the correct decision to terminate the test are issues to address prior to beginning the assessment.

Normative Data and Interpretation

Table 10.1 and the lab data collection sheet contain the normative muscular endurance data values specific to men and women. You will use these normative data to classify your muscular endurance results. For table 10.1, find your sex-specific section and age-specific row to determine your upper body muscular endurance using the terms at the top of the chart. For core endurance, compare your data to the average values on the lab data sheet and then determine if you fall close to average, below (less endurance), or above (more endurance). Additionally, calculate your ratios for flexor to extensor endurance time and lateral endurance (both right and left) to extensor endurance time to determine overall balance between these areas. These ratios are applicable and can indicate risk of low back issues based on muscular endurance imbalances (McGill 2015). Take these results and develop an overall picture of your muscular endurance. What are the ratios for the endurance tests, and do they fall within normal limits? What areas need attention to regain this balance if needed? It is important to assess in multiple areas of the body to get a full picture of overall muscular endurance. Again, one single test does not give us all the answers.

Just like in previous labs, these data should also be interpreted with a functional fitness focus in mind. Maintaining muscular endurance at all points across the lifespan is associated with improved physical function, whole body stability, and decreased risk of chronic problems like low back pain (Beltz, Nuñez, and Janot 2019; Gibson, Wagner, and Heyward 2019; Hayden, van Tulder, and Tomlinson 2005; Kell and Asmundson 2009; Mayer et al. 2015; McGill 2015; Moon et al. 2013; Nuñez et al. 2017; Rainville et al. 2004). Remember,

TABLE 10.1   Classification of Results for the YMCA Bench Press Test

MEN

Age (yr)	Well above average	Above average	Average	Below average	Well below average
18-25	49	34	26	17	5
26-35	48	30	22	16	4
36-45	41	26	20	12	2
46-55	33	21	13	8	1
56-65	28	17	10	4	0
>65	22	12	8	3	0

WOMEN

Age (yr)	Well above average	Above average	Average	Below average	Well below average
18-25	49	30	21	13	2
26-35	46	29	21	13	2
36-45	41	26	17	10	1
46-55	33	20	12	6	0
56-65	29	17	9	4	0
>65	22	12	6	2	0

Note: The values are number of repetitions completed.

Adapted by permission from A.L. Gibson, D.R. Wagner, and V.H. Heyward, *Advanced Fitness Assessment and Exercise Prescription*, 8th ed. (Champaign, IL: Human Kinetics, 2019), 171. Data from YMCA of the USA (2000).

any discussion with a client should include these talking points: (1) how their muscular endurance compares to others (classification), (2) what these data mean for their current physical function and for potential future issues (interpretation), (3) whether they should either maintain or improve their current level of endurance and in what areas (interpretation), and (4) what they should do to achieve their goals (interpretation and planning of exercise program). By focusing on these points, you will be able to make the biggest impact on your client's understanding of the data and help them plan an appropriate exercise program.

Exercise Program Design: Muscular Endurance

The main purpose of the muscular endurance prescription is to provide a gradual exercise stimulus over time to increase the muscles' capacity to perform repetitive contraction until overall fatigue (Ratamess et al. 2009). Individual goals in a muscular endurance exercise program could include increasing the ability to do activities of daily living

in the adult who is middle aged to older, improving posture and overall stability, reducing risk of chronic disease and disability, improving overall physical function across the lifespan, increasing energy expenditure for weight management, and improving recreational or competitive sport performance (Garber et al. 2011; Ratamess et al. 2009). Remember, individualizing goals is key for program adherence and regular participation in exercise.

This section will provide you with information to design a basic core endurance program for clients seeking to improve muscular endurance in the trunk muscles used for spine stabilization. Refer to lab 11 for program design guidance for improving muscular endurance in the upper and lower body musculature. Before any type of muscular endurance training, a generalized warm-up to increase blood flow and temperature within the muscle is encouraged to prepare the body for this type of training. For core-specific exercises, repeated flexion and extension movements of the spine (see lab 9), in addition to the general warm-up, are recommended before training (McGill 2015).

Frequency

Frequency of exercise refers to the number of days a client engages in endurance training per week. The recommendation for resistance training frequency, which core endurance training falls under, is at least 2 to 3 d·wk^{-1} with more sessions based on client experience and goals (ACSM 2021; Garber et al. 2011; Magyari 2018; Ratamess et al. 2009). For beginners, it is good practice to include at least 48 h of rest in between resistance training sessions (Garber et al. 2011; Magyari 2018). However, for programs focused specifically on rehabilitation and low back health, exercises performed daily may confer greater benefits (McGill 2015).

Intensity

Intensity refers to the amount of resistance used during movement. Resistance for core endurance programs can be provided using the client's body weight, medicine balls, cable machines, or other resistance implements. The resistance load for dynamic muscular endurance exercises should be such that the client can perform up to 15 to 20 repetitions until fatigue (ACSM 2021; Garber et al. 2011; Magyari 2018; Ratamess et al. 2009). The repetition maximum (RM) load will need to be determined individually for each client through trial and error until the set resistance matches the desired RM range (e.g., 8RM-10RM or 15RM-20RM). The appropriate resistance load should elicit fatigue for the last repetition or two in the RM range. With endurance training utilizing isometric holds, intensity is determined by the time that the contraction is held and the effort put into the contraction. Other repetition ranges with varying resistance loads can be used for improving muscular endurance using different types of training strategies (i.e., circuit training, high intensity functional training, etc.) in more experienced individuals, but our focus for this lab is on the beginner.

Manipulating rest intervals is another way to provide intensity in a muscular endurance program. In general, 1 to 2 min of rest between sets is sufficient to provide adequate recovery for beginner clients due to the relatively lighter resistance loads used in this type of training (Ratamess et al. 2009). Shortening the rest intervals to less than 1 min could increase the metabolic stress and intensity challenge of the training program for greater improvements in muscular endurance. This recommendation should only be applied with more experienced clients.

Volume

Volume is defined as the total number of sets performed for either a specific muscle or muscle group or an exercise per week (ACSM 2021). A minimum of two to four sets per muscle group is recommended, with a focus on high volume exercise to improve muscular endurance (Garber et al. 2011; Ratamess et al. 2009). McGill (2015) also recommends a focus on endurance rather than strength for the spine stabilizer musculature early on in programs focused on low back fitness. Fatigue-resistant core muscles are better for long-term low back health compared to high force–generating (strength) muscles. For core endurance training, the goal here is to train the musculature without getting too fatigued. Thus, adequate rest is required during this type of training.

McGill (2015) recommends the "reverse pyramid" system when training for core endurance. This works by performing an exercise for multiple sets of repetitions that decrease with each subsequent set. For instance, start with one set of five repetitions followed by rest, then one set of four repetitions followed by rest, then one set of three repetitions, and so forth, until the total repetition or set goal is reached.

Type (Exercise Selection)

Core endurance exercises can either be static (isometric) or dynamic in nature. A typical program for a client may include both static and dynamic exercises to address any type of occupational, recreational, or sport movement demands that they may have in their everyday life. Ratamess and colleagues (2009) recommend including exercises that involve multi- and single-joint movements in a muscular endurance program. This recommendation is also supported by others (Garber et al. 2011; Gibson, Wagner, and Heyward 2019; Magyari 2018). Ultimately, the choice of exercise will be dictated by the program goals and needs of the client.

Static core exercises are done to build endurance of the spine stabilizer (core) musculature while simultaneously restricting movement in the spine

(McGill 2015). This type of training is very important for learning spine stiffening technique (i.e., abdominal bracing in neutral spine position) to protect the spine during dynamic movement and to decrease external loading of the spine during exercise while building greater endurance capacity. These two factors are especially critical for those clients with a history of past or current low back issues and poor core stability. According to McGill (2015), proper curl-up (anterior stabilizers; rectus abdominis), side bridge variations (lateral stabilizers; quadratus lumborum, obliques, transverse abdominis), and bird dog variations (posterior stabilizers; back extensors) are three exercises that should be at the center of any core endurance program addressing all spine stabilizers. Other static exercises for the core are reverse glute and other body bridges, front plank, stir the pot, and dead bug variations, among others. Remember, these exercises should be done with as limited spine motion as possible. A final piece regarding static core exercises is the length of time each isometric contraction is held. McGill (2015) recommends no more than 7 or 8 s of contraction hold per exercise followed by a short rest period (~5 s). This contraction can be repeated for as often as is required to attain the overall volume goal for a particular exercise. For example, if you want a client to perform two sets of the bird dog and front plank exercises for 1 min each set, then both exercises should be broken up into multiple repetitions of 7 or 8 s holds that equal the total volume goal. Thus, continuous plank or other isometric holds for minutes on end should not be recommended!

Dynamic core exercises can include multidirectional movements involving rotation in the transverse plane, flexion and extension in the sagittal plane, abduction and adduction in the frontal plane, or in a diagonal pattern (Magyari 2018; McGill 2015). Modalities that can be used are cable machines with a weight stack to perform horizontal cable chops and high to low or low to high cable chops; medicine balls to perform rotational throws, slams, reverse throws, and presses; and dynamic balance exercises with a focus on core stability. Also, "anti-movement" exercises for the core such as "anti-rotation," "anti-lateral flexion," and "anti-extension" Pallof-type presses and cable walkouts can be done using cables or bands.

Additionally, performing these and other resistance training exercises utilizing an unstable surface (e.g., stability ball, foam pad, BOSU, etc.) or modality (e.g., TRX training) can provide both a dynamic core and balance challenge for the client (Behm and Colado 2012; Behm and Sanchez 2013; Behm et al. 2015; Janot et al. 2013; Magyari 2018). However, if greater force or power production is the goal, it is recommended that a more stable surface is used during exercise (Behm and Anderson 2006). Also, employing unstable surfaces during exercise leads to greater co-contraction of the spine stabilizing muscles, which equally increases the stress or load on the spine (McGill 2015). Therefore, unstable surfaces for certain clients should be prescribed cautiously, and that client should be carefully assessed first, especially if low back health is an issue.

Finally, velocity of contraction during muscular endurance training can vary along a large continuum of tempos and still be effective. It is generally recommended that moderate to faster velocities (i.e., a ratio of 2 s during concentric movement and 2 s during eccentric movement or less) of contraction per repetition be performed during muscular endurance training (Ratamess et al. 2009). With that said, it is imperative that form is never sacrificed for greater velocity; thus, beginners should perform at the tempo at which they can best maintain technique and form during exercise.

Progression

Progression refers to changes in training stimuli in order to maintain training improvements over time (ACSM 2021). Progression in a core endurance program can be done by altering the volume of exercise, intensity (resistance), and the type of exercises (ACSM 2021). In general, if volume is increased, resistance is kept constant or decreased to balance out the overload. In contrast, when intensity is increased, volume is kept constant or decreased to balance the overload.

Volume can be progressed by increasing training frequency, number of sets, and number of repetitions. If you choose to increase volume, one factor to keep in mind is time per exercise session. Time can be an issue affecting adherence and therefore should be planned according to the client's time availability and overall goals.

Intensity is mostly progressed through increases in resistance load. A general rule of thumb is to increase resistance to maintain the RM load, meaning that the prescribed resistance should elicit fatigue within the last repetition or two for

Case Study for Core Muscular Endurance Exercise Prescription

For the program design, you will fill in the core endurance exercise program planning sheet with your recommendations for type of exercise, number of sets and repetitions performed for each exercise, resistance, rest periods between repetitions, and rating of perceived exertion (1-10 scale). For simplicity, this program is for a client who is new to core endurance exercise and at the initial stage of their program. You can choose up to 10 different exercises addressing muscles involved in each area of the trunk (i.e., anterior, lateral, and posterior). These exercises do not need to be done every session; therefore, you have the freedom to choose on which days they will be done. For examples of select core muscular endurance exercises, review appendix F.3 in the *Advanced Fitness Assessment and Exercise Prescription* text by Gibson, Wagner, and Heyward (2019) or the *Low Back Disorders: Evidence-Based Prevention and Rehabilitation* text by McGill (2015).

In the absence of working with a real client, do your best to apply the muscular endurance program design principles through this case study example. When you are finished, turn in your work to get feedback on your overall plan.

Demography

Age: 34
Height: 5 ft 9 in. (175.3 cm)
Weight: 180 lb (81.8 kg)
Sex: Male
Race or Ethnicity: White

Family History

Your client's family history reveals cardiovascular disease on their father's side (their paternal grandfather and uncle died of myocardial infarction). No other significant issues were reported.

Medical History

Present Conditions

On the health history questionnaire form, your client reports transient, moderate pain and discomfort in the lower back. This occurs mostly during exercise and at home following various activities that involve lifting. Your client does not take any pain relievers to manage the pain. A recent physical assessment revealed the following health information. Your client's resting HR was 75 beats·min^{-1} and resting BP was 116/72 mmHg (confirmed from a previous measurement). The client's skinfold assessment showed a percent body fat (%BF) of 20%. No blood chemistry panel for cholesterol or blood glucose is reported at this time. Their current exercise test information reveals a $\dot{V}O_2$max value of 46 mL·kg^{-1}·min^{-1} (submaximal running test), a 1.5 and 2.1 strength-to-mass ratio value for the 1RM bench press and 1RM leg press tests, respectively, a grip strength of 110 kg (242 lb), and true lumbar flexion value of 41 degrees. Following is the muscular endurance information you collected. No other data are available currently.

YMCA bench press test	Score (reps)	Rating	Push-up test	Score (reps)	Rating
Upper body	35		Upper body	28	
Core endurance test	Score (s)	Rating	Core endurance test	Score (s)	Rating
Flexor	240		Lateral (right) SB	93	
Extensor	91		Lateral (left) SB	83	
Flexor/Extensor ratio	2.64		Right SB/Extension ratio	1.02	
			Left SB/Extension ratio	0.91	

Note: SB = side bridge.

> continued

Case Study for Core Muscular Endurance Exercise Prescription   > continued

Past Conditions

Your client reports two surgeries, one on their abdominal area for a sports hernia (10 yr ago) and one on the right shoulder for rotator cuff repair (2 yr ago). They have sought physical therapy services over the years for knee, shoulder, elbow, and hip issues. Otherwise, your client reports no other medical and health issues.

Behavior and Risk Assessment

Your client is a nonsmoker who performs regular physical activity. They have a semi-active job as a university professor that includes periods of sedentary behavior combined with low intensity movement (i.e., standing and walking) throughout the day. Sitting for long periods of time and certain movements, mostly during resistance training, can trigger pain and discomfort in the low back and hips. They are currently in the last week of physical therapy to correct deficient movement patterns, improve motor control of muscles around the hip and spine, and learn spine stabilizing techniques during movement such as abdominal bracing. Your client would like to begin a structured core endurance exercise program to accompany the 16 wk cardiorespiratory fitness, flexibility, and strength program already in place. The program that you have designed includes 4 d·wk^{-1} of moderate to vigorous intensity physical activity, mostly uphill treadmill walking or outside at a brisk pace for 30 to 40 min·d^{-1}, and 2 d·wk^{-1} of moderate intensity resistance training at the university fitness center. Flexibility training will be done every day at a set time in the evening. The main goals for the core endurance program are to improve their core endurance test scores to above average where indicated, have better balance between flexor and extensor musculature endurance (ratio ≤1.0), and avoid low back pain and discomfort during activity. They would like to fit core endurance training in following resistance training on those days and at least one other day during the week. There is no information collected on your client's diet.

Review your client's data and provide a rating for each of the assessments in the data table. Also, please fill out the program planning sheet with exercises and other information for your client to follow based on the exercise goals.

each exercise. If your client reports that the last few repetitions within the RM range are easy or less challenging, it may be time to increase the load.

Exercise selection allows for the greatest variety and creativity as a progression strategy. Exercises can be progressed based on familiarity, such as beginning a program with exercises that a client is familiar with and progressing to exercises that the client has less experience with. Progression can also be done based on exercise complexity, such as beginning a program with exercises that are easier to master and require less skill or fitness to perform and progressing to exercises that are more difficult, require more effort, and are more complex from a movement perspective.

Familiarity and complexity can be manipulated within the same exercise or across different exercises. For instance, the bird dog exercise, done to challenge the back extensors, could begin with simple single arm and leg extension holds and progress to more complex movements such as the contralateral bird dog (i.e., opposite arm and opposite leg are lifted) or the contralateral bird dog with the arm and leg moving in geometric patterns while the client stabilizes their core. A familiar and simple exercise like the anti-rotation Pallof press using a cable and weight stack can be progressed to a medicine ball rotational toss, which is a dynamic exercise requiring whole body joint stabilization and a higher velocity of contraction in the core to perform.

The exercise professional can utilize exercise selection to enhance variety to reduce potential boredom and staleness. To further increase adherence, ensure that your client enjoys the exercises, determine if they are appropriately challenged and not at risk for injury or overtraining, and always evaluate whether or not the exercises are related to the overall program goals.

MUSCULAR ENDURANCE DATA SHEET

Name _____ Sex _____ Date _____

Age _____   Height _____cm   Weight _____kg

YMCA Bench Press Test

| Weight used _____ lb or kg | Number of repetitions _____ reps |

Note: Set the metronome to 60 beats·min⁻¹. Men use 80 lb (36.4 kg) and women use 35 lb (15.9 kg).

Percentile rank and rating from table 10.1 _____

Core Endurance Test Battery

Flexor endurance test		Average scores for men and women	
Time until failure _____ s	Rating _____	Men: 136 s	Women: 134 s
Extensor endurance test		Average scores for men and women	
Time until failure _____ s	Rating _____	Men: 161 s	Women: 185 s
Lateral endurance test: Right side bridge		Average scores for men and women	
Time until failure _____ s	Rating _____	Men: 95 s	Women: 75 s
Lateral endurance test: Left side bridge		Average scores for men and women	
Time until failure _____ s	Rating _____	Men: 99 s	Women: 78 s

Flexor/extensor endurance ratio _____ Rating _____
This value should be ≤1.0. A test result >1.0 is indicative of poor balance among these stabilizers.

Lateral endurance (RSB)/extension ratio _____ Rating _____
Lateral endurance (LSB)/extension ratio _____ Rating _____
These values should be ≤0.75. A test result >0.75 is indicative of poor balance among these stabilizers.

Note: RSB = right side bridge; LSB = left side bridge.
Data from McGill (2015, p. 288).

From J. Janot and N. Beltz, *Laboratory Assessment and Exercise Prescription* (Champaign, IL: Human Kinetics, 2023).

CORE ENDURANCE EXERCISE PROGRAM PLANNING SHEET

Frequency _____ d·wk^{-1}

Exercise	Day	Sets	Repetitions	Resistance	Rest (s)	RPE

Note: Choose your exercises based on your client's goals. These do not have to be done every workout and can be planned over multiple days to fit all these exercises into the program.

From J. Janot and N. Beltz, *Laboratory Assessment and Exercise Prescription* (Champaign, IL: Human Kinetics, 2023).

Assessment of Muscular Fitness: Muscular Strength

Purpose of the Lab

As we briefly addressed in lab 2, an adequate level of muscular strength is needed for a broad range of activities throughout the lifespan. From fundamental movement patterns during early development to complex athletic-based tasks and activities of daily living, muscular strength underpins the ability to generate force across an array of time constraints. Muscular strength development and the preservation of muscle mass is often regarded as a primary goal for individuals who are young and healthy and those with athletic aspirations, but muscular strength is also imperative for balance, fall prevention, and independence throughout the aging process (Pizzigalli et al. 2011). The age-related decline in skeletal muscle mass, muscular strength, and overall physical performance is known as sarcopenia (Rosenberg 1997). It is estimated that the prevalence of sarcopenia is 6% to 12% worldwide (Shimokata et al. 2018). This multifaceted decline not only increases the risk for chronic diseases such as osteoporosis, cardiovascular disease, dementia, respiratory disease, and type 2 diabetes but also contributes to diminishing quality of life (Pacifico et al. 2020).

The most effective means to increase muscular strength, preserve muscle mass, and combat the onset and progression of sarcopenia is regular resistance training (RT) exercise. Skeletal muscle can exhibit many types of muscle actions. In lab 2, we tested isometric muscular contraction with grip strength. The muscle was static during force production, neither shortening nor lengthening. Shortening of the muscle during force production is known as concentric muscle action, and lengthening of the muscle is eccentric action. Additionally, we can test force production while controlling the speed at which the joint moves through a range of motion (ROM). This type of muscle action is known as isokinetic, and testing requires an isokinetic testing system. Isokinetic testing will not be covered in detail throughout this lab activity because it is more common in a graduate-level or advanced laboratory setting. Since most human tasks involving skeletal muscle force production are not completed in a static or isokinetic environment, the most appropriate way to evaluate strength is though dynamic assessment. We evaluate muscular strength to obtain a baseline prior to starting an RT program, classify level of strength, identify any imbalances or weak areas that need improvement, track progress, and use maximal values to prescribe RT.

In this lab, you will learn how to administer one-repetition maximum (1RM) and estimated 1RM testing to assess dynamic muscular strength using free weights and machines. Using the information from maximal strength testing, you will evaluate relative upper and lower body muscular strength and then use that information to develop an individualized RT prescription.

Necessary Equipment or Materials

- Body weight scale
- Stopwatch
- Bench press with barbell and safety clips
- Leg press machine
- Weight plates in a variety of sizes
- Muscular strength assessment data sheet
- Resistance training program planning sheet

Calibration of Equipment

For this lab, no calibration of equipment is needed. Make sure that all equipment is in good working condition before use.

Procedures

One-repetition maximum testing is a common assessment to measure dynamic muscular strength. It is safe to complete 1RM testing on individuals if they have been instructed on proper technique and are being spotted throughout the testing process. If your client has an underlying condition that requires a weight limit restriction, maximal strength testing may not be recommended. If you determine that a client is not skilled or comfortable enough to complete a true 1RM safely, then an estimated 1RM test can be completed. The goal of

1RM testing is to identify the maximum load that an individual can move exactly one time through trial and error. Estimated 1RM testing is completed using less weight to elicit failure between two and five repetitions. The total weight lifted and the total repetitions completed can be used in a prediction equation to estimate 1RM. A spotter should be used when completing 1RM or estimated 1RM testing. Always follow the recommended client positioning and spotter guidelines for each exercise completed in this lab. The upper body muscular strength assessment completed in this lab is the barbell bench press, and the lower body assessment is the leg press. The same 1RM or estimated 1RM procedure will be followed for both assessments.

1RM Protocol (Haff and Triplett 2016)

1. Have the client warm up with a light amount of weight for 5 to 10 repetitions. There should not be any struggle in completing this first warm-up set.
2. Allow a 1 min rest period.
3. Complete another warm-up set of three to five repetitions by adding 5% to 10% more weight for bench press and 10% to 20% more weight for leg press.
4. Allow a 2 min rest period.
5. Complete a near-maximal warm-up set of two to three repetitions by adding 5% to 10% more weight for bench press and 10% to 20% more weight for leg press.
6. Allow a 2 to 4 min rest period.
7. Add 5% to 10% more weight for bench press and 10% to 20% more weight for leg press.
8. Instruct the client to attempt a single repetition.
   a. If the client successfully completes the repetition, allow a 2 to 4 min rest period. Complete step 7 again.
   b. If the client fails to complete the single repetition, allow a 2 to 4 min rest period. Decrease the weight by 2.5% to 5% for bench press and 5% to 10% for leg press and complete step 8 again.
9. Continue with the testing process, adding or removing weight until the client establishes a 1RM value with proper technique.

10. Record all information on the lab 11 data collection sheet.

11. Calculate relative strength by dividing 1RM by the client's body mass.

    Relative Strength = 1RM (lb) / body mass (lb)

12. Interpret maximal muscular strength using table 11.1 or 11.2.

Estimated 1RM Protocol

1. Have the client warm up with a light amount of weight for 5 to 10 repetitions. There should not be any struggle in completing this first warm-up set.

2. Allow a 1 min rest period.

3. Complete another warm-up set of three to five repetitions by adding 2.5% to 5% more weight for bench press and 5% to 10% more weight for leg press.

4. Allow a 2 min rest period.

5. Complete a third warm-up set of two to three repetitions by adding 2.5% to 5% more weight for bench press and 5% to 10% more weight for leg press.

6. Allow a 2 to 4 min rest period.

7. Add 2.5% to 5% more weight for bench press and 5% to 10% more weight for leg press.

8. Instruct the client to attempt as many repetitions as possible.

   a. If the client successfully reaches five repetitions, instruct the client to stop and allow a 2 to 4 min rest period. Complete step 7 again.

   b. If the client completes a single repetition, go to step 10 in the 1RM Protocol section.

   c. If the client fails to complete a single repetition, allow a 2 to 4 min rest period. Decrease the weight by 2.5% for bench press and 5% for leg press and complete step 8 again.

9. Continue with the testing process, adding or removing weight until the client successfully completes a set between two and five repetitions.

10. Record all information on the lab 11 data collection sheet.

11. Estimate 1RM by using one of the following equations:

    a. For successful completion of two to four repetitions (Brzycki 1993):

       1RM = weight lifted (lb) / [1.0278 − (reps completed × 0.0278)]

    b. For successful completion of five repetitions (Reynolds, Gordon, and Robergs 2006):

       Bench Press: 1RM = (1.1307 × weight lifted [kg]) + 0.6998

       Leg Press: 1RM = (1.09703 × weight lifted [kg]) + 14.2546

12. Calculate relative strength by dividing 1RM by the client's body mass.

    Relative Strength = 1RM (lb) / body mass (lb)

13. Interpret maximal muscular strength using table 11.1 or 11.2.

Bench Press Technique (Haff and Triplett 2016)

1. Lie on the bench in a supine position with the head, shoulders, and buttocks contacting the bench and feet flat on the floor. Align the head so the barbell is directly above the eyes.

2. Grasp the bar tightly with an overhand grip and hands slightly wider apart than the shoulders.

3. With assistance from the spotter, lift the bar off the rack by extending the elbows.

4. Lower the bar at the nipple and contact the chest.

5. Press the bar away from the chest until the elbows are fully extended once again.

6. Be sure to maintain contact with the bench throughout the entire movement and avoid back arching. Always use clips to secure the plates on the barbell.

Spotting Technique

1. Stand directly behind the head of the client with feet shoulder-width apart.

2. Grasp the bar inside the client's hands using an alternating grip.

3. Assist the client on the initial lift from racked position, as requested.

4. Without touching the bar, guide your hands along with the bar during the downward and upward phases of the lift.

TABLE 11.1   Norms for Relative One-Repetition Maximum Bench Press Strength by Age and Sex

Age (yr)	PERCENTILE								
	10	20	30	40	50	60	70	80	90
MEN									
20-29	0.80	0.88	0.93	0.99	1.06	1.14	1.22	1.32	1.48
30-39	0.71	0.78	0.83	0.88	0.93	0.98	1.04	1.12	1.24
40-49	0.65	0.72	0.76	0.80	0.84	0.88	0.93	1.00	1.10
50-59	0.57	0.63	0.68	0.71	0.75	0.79	0.84	0.90	0.97
60+	0.53	0.57	0.63	0.66	0.68	0.72	0.77	0.82	0.89
WOMEN									
20-29	0.30	0.33	0.35	0.37	0.40	0.41	0.42	0.49	0.54
30-39	0.27	0.32	0.34	0.37	0.38	0.41	0.42	0.45	0.49
40-49	0.23	0.27	0.30	0.32	0.34	0.37	0.38	0.40	0.49
50-59	0.19	0.23	0.26	0.28	0.31	0.33	0.35	0.37	0.40
60-69	0.25	0.26	0.28	0.29	0.30	0.32	0.36	0.38	0.41
70+	0.20	0.21	0.24	0.25	0.27	0.31	0.33	0.39	0.44

Note: Relative strength = 1RM / body mass.

Adapted by permission from A.L. Gibson, D.R. Wagner, and V.H. Heyward, *Advanced Fitness Assessment and Exercise Prescription*, 8th ed. (Champaign, IL: Human Kinetics, 2019), 168. Data for women provided by the Women's Exercise Research Center, The George Washington University Medical Center, Washington, DC, 1998. Data for men provided by The Cooper Institute for Aerobics Research, *The Physical Fitness Specialist Manual*, The Cooper Institute, Dallas, TX, 2005.

TABLE 11.2   Norms for Relative One-Repetition Maximum Leg Press Strength by Age and Sex

Age (yr)	PERCENTILE								
	10	20	30	40	50	60	70	80	90
MEN									
20-29	1.51	1.63	1.74	1.83	1.91	1.97	2.05	2.13	2.27
30-39	1.43	1.52	1.59	1.65	1.71	1.77	1.85	1.93	2.07
40-49	1.35	1.44	1.51	1.57	1.62	1.68	1.74	1.82	1.92
50-59	1.22	1.32	1.39	1.46	1.52	1.58	1.64	1.71	1.80
60+	1.16	1.25	1.30	1.38	1.43	1.49	1.56	1.62	1.73
WOMEN									
20-29	1.02	1.13	1.23	1.25	1.32	1.36	1.42	1.66	2.05
30-39	0.94	1.09	1.16	1.21	1.26	1.32	1.47	1.50	1.73
40-49	0.76	0.94	1.03	1.12	1.19	1.26	1.35	1.46	1.63
50-59	0.75	0.86	0.95	1.03	1.09	1.18	1.24	1.30	1.51
60-69	0.84	0.94	0.98	1.04	1.08	1.15	1.18	1.25	1.40
70+	0.75	0.79	0.82	0.83	0.89	0.95	1.10	1.12	1.27

Note: Relative strength = 1RM / body mass.

Adapted by permission from A.L. Gibson, D.R. Wagner, and V.H. Heyward, *Advanced Fitness Assessment and Exercise Prescription*, 8th ed. (Champaign, IL: Human Kinetics, 2019), 169. Data for women provided by the Women's Exercise Research Center, The George Washington University Medical Center, Washington, DC, 1998. Data for men provided by The Cooper Institute for Aerobics Research, *The Physical Fitness Specialist Manual*, The Cooper Institute, Dallas, TX, 2005.

5. Keep proper posture during the movement phase by bending at the knees and keeping the back flat.

6. When necessary, help prevent the bar from descending back toward the client if muscular failure is reached.

7. Grasp the bar again and slowly place it back into the racked position when the desired number of repetitions have been completed.

Leg Press Technique (Haff and Triplett 2016)

1. Sit in the leg press machine and place the feet on the platform, hip-width apart with toes pointing slightly outward.

2. Keep the legs in parallel, grasp the handles on the sides of the machine, and extend the knees to press against the platform. Release the platform support mechanism.

3. Keeping the feet, knees, and hips aligned, allow the platform to slowly lower.

4. Always keep the back, hips, and buttocks against the back pad of the machine. Do not allow the feet to shift or lift during movement.

5. Allow flexion of the knees and hips until the mid-thighs are parallel to the platform.

6. Push against the platform until the knees are in an extended position, but do not forcibly lock the knees.

7. Once the desired number of repetitions have been completed and the platform is in the fully extended position, grasp the side handles and apply the safety mechanism.

Spotting Technique

The leg press machine has a safety mechanism built in to assist clients, but spotting may be necessary when performing 1RM testing. It is particularly helpful when the platform cannot be moved from the bottom position.

1. Stand to the side of the leg press machine and grasp the plate loading bar on the platform with both hands.

2. Without providing any assistance, guide the platform with your hands as it is moved upward and downward.

3. Be sure to bend with the knees and hips as the platform moves along its path, and avoid bending at the back.

4. Allow the client to provide the resistance during successful repetitions. When maximal exertion or failure is reached, assist until the platform can reach its closest mechanical assist stopping position.

Common Errors and Assumptions

One-repetition maximum testing assumes a linear relationship between the weight lifted and the number of repetitions that can be successfully completed to exhaustion; however, the actual relationship is more curvilinear (LeSuer et al. 1997; Mayhew, Ware, and Prinster 1993). Therefore, any increase in repetitions to failure during maximal strength testing increases the error in a 1RM prediction equation or chart. The main source of error during testing is the client's exercise technique and overall effort. The sources are similar, as they relate to your client's level of experience and familiarity with RT. You will need to evaluate your client's technique prior to testing, observing them during the warm-up sequence and providing verbal encouragement during testing to ensure maximal effort. This is important for both client safety and the efficacy of testing.

Normative Data and Interpretation

Interpret your bench press and leg press results using tables 11.1 and 11.2. Remember to explain your interpretations in a way that makes them understandable and significant for your client. For example, if your 32-year-old male client weighs 197 lb (89.4 kg) and completes a 1RM leg press of 355 lb (161 kg), the relative leg press strength is 1.80. A relative leg press strength score of 1.80 puts them between the 60th and 70th percentile. This interpretation may not be understandable or significant to all clients. Rephrasing it as "your leg strength is higher than 60 to 70 percent of males your age" or "only 30 to 40 percent of men your age have greater leg strength" will deliver greater understanding and significance.

Exercise Program Design: Resistance Training Exercise Prescription

Although multiple forms of resistance training are available (isokinetic, isometric, variable resistance) we will focus on dynamic RT in this lab activity. This lab will also present multiple perspectives on prescribing dynamic RT. Regular participation in an RT program can accomplish a myriad of goals. Namely, RT can increase or maintain muscular strength, muscular endurance, and skeletal muscle mass and improve physical performance, resting blood pressure, insulin resistance and glycemic control, bone mineral density, resting metabolic rate, balance, functional independence, cognitive abilities, and self-esteem (Westcott 2012). We recommend a short dynamic warm-up prior to an RT session, comprised of the movement patterns to be trained during that specific day. An aerobic cooldown, static stretching session, or aerobic exercise session can certainly be included before or after an RT session. The considerations for an RT program according to the American College of Sports Medicine (2021a) guidelines are (1) frequency, (2) intensity, (3) type, (4) rest intervals, (5) volume (sets and repetitions), and (6) progression. Additional components of RT design are the needs analysis and exercise order considerations, addressed in greater detail by the National Strength and Conditioning Association (NSCA) (Haff and Triplett 2016). These parts of an individual RT prescription will be discussed throughout the remainder of this lab.

Needs Analysis

The needs analysis portion of the RT design focuses on the evaluation of the client's common daily task performance, their RT and health history, and any individual training goals. A sport performance task is an easily understood concept when conducting the needs analysis. For example, if you are creating an RT program for a high school baseball pitcher, you would parse the individual phases of a baseball pitch and identify the muscle groups necessary for those movement patterns. This same concept can be applied to the general population or clinical client. For example, you can list the activities of daily living (ADLs) typically completed by your clinical client and identify the common movement patterns associated with those ADLs as well as the muscle groups and actions required

to complete those tasks successfully. You will also need to discuss RT background and health history with your client. This will help you understand the capabilities of your client and guide you in the process of selecting specific exercises. For example, if your client is 57 years old and last engaged in RT during high school, a barbell back squat may not be the most appropriate exercise selection because it requires a high level of technique and underlying core stability to complete safely. Discussion about health history, particularly anything that may preclude safe completion of specific exercises, such as past or present injuries, will also assist you in appropriate exercise selection. Finally, your client should have a specific goal relating to RT. They may want to increase overall muscular fitness (strength, endurance, power, flexibility) or have a specific goal within those components. Goals may also relate to participation in specific tasks, such as wanting to engage in recreational activities.

Frequency

Training frequency refers to the number of days in which a client participates in RT per week. The needs analysis step in program design will help you determine the appropriate frequency for your client. The ACSM (2021a) recommends engaging in RT at least 2 d per week for individuals without previous RT experience while the NSCA (Haff and Triplett 2016) recommend 2 to 3 d per week for beginners. If your client has time and experience constraints, it may be appropriate to start with 1 day per week and progress to a higher frequency as their restrictions subside. The frequency can also depend on the design of the routine. For example, a split routine focuses on specific movement patterns or muscle groups during a single RT session. This allows for a greater training frequency throughout the week because appropriate rest days are given to muscle groups. We will focus on a full body RT design in this lab. Generally, it is recommended to complete full body RT sessions on nonconsecutive days. This means that the highest training frequency is 3 to 4 d per week. Keep in mind that the intensity of the RT session may also affect the frequency, where more intense training days may require more than one rest day between sessions.

Intensity (Load)

Intensity or load refers to the amount of weight being lifted during an exercise. The most precise way to prescribe RT intensity is using %1RM. Using the results from the 1RM testing completed in this lab, we can assign a load based on the specific training goals of the client. The assigned load also determines the repetition goal for a given set of an exercise, where higher repetitions can be completed at lower intensities and vice versa. Currently, the ACSM (2021a) recommends an intensity of 60 to 70 %1RM and a repetition range of 8 to 12 to improve muscular fitness. Specific muscular fitness goals have different recommendations for load and repetition range. The ACSM (2021a) recommends an intensity of >80 %1RM and repetition range between one and six repetitions for muscular strength development, although lower intensities (40-85 %1RM) have been shown to elicit strength gains in untrained individuals (Peterson, Rhea, and Alvar 2005). The NSCA recommends a similar intensity (≥85 %1RM) and repetition range (≤6) for optimal muscular strength development. Increasing (hypertrophy) or preserving muscle mass is best accomplished with loads between 67 and 85 %1RM and a repetition range of 6 to 12 (ACSM 2021a; Haff and Triplett 2016). Regarding muscular endurance, the ACSM (2021a) generally recommends lower loads and higher repetition ranges while the NSCA (Haff and Triplett 2016) gives an optimal load of ≥67 %1RM and repetition range of ≥12. It is important to note that 1RM testing every single exercise is not feasible, presenting a challenge to load manipulation. For example, controlling the load for a 1RM test on a dumbbell walking lunge is difficult to accomplish and unnecessary. In addition, using near-maximal loads may be contraindicated altogether with your client. An effective alternative to prescribing load using %1RM is using a rating of perceived exertion (CR10) scale to indicate repetitions in reserve (RIR). While the RIR method may not have the precision of the %1RM method, it considers day-to-day variability in the client (stress, sleep, residual soreness) that affects successful completion of set goals (Helms et al. 2016). For example, a client may complete a set of leg presses at 75 %1RM for 10 repetitions on a Monday but fail on the seventh repetition on the following Thursday. The RIR method uses the CR10 scale to indicate the number of repetitions that a client could have completed during the set. A description of the CR10 values and corresponding RIR prompt are shown in table 11.3.

Recommendations using the RIR method for each muscular fitness goal are the following:

TABLE 11.3 Rating of Perceived Exertion (CR10) and Repetitions in Reserve Equivalents

CR10 rating	Repetitions in reserve
1-4	Little to light effort
5-7	3-6 repetitions remaining
8-9	1-2 repetitions remaining
10	Maximum effort

Adapted from Zourdos et al. (2016).

Muscular strength: 1-6 repetitions at CR10 = 7-10 or 0-3 RIR

Muscular hypertrophy: 6-12 repetitions at CR10 = 8-10 or 0-2 RIR

Muscular endurance: 12+ repetitions at CR10 = 9-10 or 0-1 RIR

We know that lower repetition ranges warrant greater resistance loads, but the principle that you'll notice immediately when using the RIR method is that all muscular fitness adaptations happen in response to near-maximal effort on every set of an exercise.

The amount of rest taken between sets of exercises is dependent upon the goal of the program and the intensity of the set. Sets that have higher loads will require more rest than those with lower weight. Rest period recommendations for each training goal are as follows (Haff and Triplett 2016):

Muscular strength: 2-5 min

Muscular hypertrophy: 30-90 s

Muscular endurance: ≤30 s

Volume

Training volume refers to the entire load lifted during an RT session. The number of exercises, the number of sets for each exercise, and the repetitions within each set all comprise the training volume. The ACSM (2021a) recommends two to four sets of each exercise per RT session. Individuals with limited RT experience should start by performing two sets of each exercise and progress to four sets as they become better trained. The NSCA (Haff and Triplett 2016) set recommendations based on training goals are as follows:

Muscular strength: 2-6 sets

Muscular hypertrophy: 3-6 sets

Muscular endurance: 2-3 sets

Type (Exercise Selection)

The ACSM (2021a) currently recommends the use of free weights, body weight, machines, and resistance band exercises to improve muscular fitness. These modalities should be used to design a program consisting of multi-joint, single-joint, and core exercises. Furthermore, these exercises should emphasize all three main muscle actions (concentric, eccentric, isometric) and target both agonist and antagonist muscle groups to avoid imbalances. The NSCA (Haff and Triplett 2016) recommends these elements but also places emphasis on designing programs based on results of the needs analysis (i.e., RT experience, movement analysis). Traditionally, exercises are selected to target all major muscle groups, such as shoulders, chest, arms, trunk, and legs. The main issue with this approach is that it is often dominated with sagittal plane movement patterns rather than focusing on the multiplanar patterns exhibited in everyday activities (Whitehurst et al. 2005). For this reason, we recommend selecting exercises based on a balanced multiplanar approach. When building a planar-based program, emphasize multiplanar joints and progress to single-plane joints.

Exercise order is also a consideration once you have selected the exercises to be completed during a session. Exercise order is the sequence in which exercises are performed during a single RT session. An appropriate exercise order prevents unnecessary fatigue to specific muscle groups. For example, performing pull-ups after completing biceps curls may affect the success of the pull-up set because the biceps are an accessory muscle during the pull-up exercise. This can be the case even when adequate rest periods are given. A few strategies exist to give flexibility in designing exercise order. One method is alternating pushing and pulling exercises. This strategy can be used when performing consecutive exercises in the same upper or lower body group. For example, you can alternate a hamstring curl

---

Case Study for Resistance Exercise Prescription

For the program design, you will fill in the resistance training program planning sheet with your recommendations for each exercise, intensity, number of sets and repetitions performed for each exercise, and rest periods between sets. This prescription is for a client who is new to resistance training exercise and at the initial stage of their program. They will be engaging in a full body program 3 d per week. Complete your planar chart with 20 different exercises, accomplishing all joints and planes of motion. Mix and match those exercises to design a 3 d per week RT program. For examples of RT exercises, review the *Essentials of Strength Training and Conditioning, Fourth Edition* text by Haff and Triplett (2016).

Demography

Age: 62
Height: 5 ft 3 in. (160.1 cm)
Weight: 140.1 lb (63.7 kg)
Sex: Female
Race or Ethnicity: White

Family History

Your client's brother died of a heart attack at age 46 and their sister passed away at 57 from complications resulting from type 2 diabetes. Their mother passed away at age 81 from a heart attack and their father passed away at age 61 from liver cancer.

Medical History

Present Conditions

On the health history questionnaire form, your client reports mild pain, discomfort, and limited range of motion in the middle and lower back. The onset of pain is consistent and worsens as the day progresses. A recent physical assessment revealed the following health information. Your client's resting HR was 71 beats·min⁻¹ and resting BP was 110/74 mmHg (confirmed from a previous

and leg extension or a bench press and seated row consecutively during an RT session. Another popular strategy is alternating upper and lower body exercises altogether. The upper–lower body alternating strategy may also allow for shorter rest periods during RT sessions with time constraints. Where the push–pull alternating strategy may fatigue similar stabilizing or accessory muscles, the upper–lower body alternating strategy will avoid this altogether.

Progression

An individual will have adaptations in their muscular fitness throughout their program. It is important to manipulate the RT variables so they may continue to meet their goals rather than plateau in their program. Providing additional stress and challenge in the program is necessary to elicit further adaptations, but it must be done systematically to avoid overtraining. One strategy to guide progression is known as the 2-for-2 rule (Baechle and Earle 2020). This rule states that additional weight can be added to a particular exercise if an individual is able to complete two or more repetitions above their assigned repetition goal during their last set for two consecutive workouts. The general recommendation for intensity progression is a load increase of 5% to 10% per week (ACSM 2021b). Conceptually, this works well with an RIR design if a client reports a lower CR10 after completing the final set of the exercise for that workout. For example, if a client completes two sets of the leg press with a goal CR10 of 8 (2 RIR) but they report a 6 (4 RIR) for two consecutive workouts, it is appropriate to add weight to the leg press for the next RT session. Another option for progression is to increase the goal repetitions within the set of an exercise if it stays within the repetition range for the muscular fitness goal. Using the leg press example, we could progress the repetition goal from 10 to 12 repetitions while maintaining the same weight if the client had a muscular hypertrophy goal.

measurement). The client's BIA showed a percent body fat of 29.6%. Fasting blood glucose is 92 $mg \cdot dL^{-1}$. They are currently taking atorvastatin to help lower their total cholesterol. Their current exercise test, which was a Rockport 1-mile walk test, showed a $\dot{V}O_2max$ value of 33.9 $mL \cdot kg^{-1} \cdot min^{-1}$. The muscular strength testing data that you collected are as follows.

Estimated 1RM leg press: 145 lb (65.8 kg) Relative strength score _____ Rating _____	Estimated 1RM bench press: 65 lb (29.5 kg) Relative strength score _____ Rating _____
Grip strength test Right: 23 kg, 27 kg, 24 kg Left: 21 kg, 20 kg, 23 kg Sum of best L + R _____ kg Rating _____	

Past Conditions

Your client was in a car accident 15 yr ago that led to accelerated thoracic and lumbar disc degeneration. They had transforaminal lumbar interbody fusion surgery approximately 2 yr ago to fuse the L1–L3 vertebrae.

Behavior and Risk Assessment

Your client has been physically active at least 4 $d \cdot wk^{-1}$ for the past 10 yr. Most of their activity consists of walking and cycling for 30 to 45 min per session. They have been retired for 3 yr. Your client is coming to you to develop a structured resistance exercise program to accompany the core endurance and flexibility program that you have already made for her. The program that you have designed includes 3 $d \cdot wk^{-1}$ of resistance training at the local university gym for the 15 wk summer session. Their main goals for the resistance training program are to increase core and lower body strength. There is no information collected on your client's diet.

Review your client's data and provide a rating for each of the assessments in the data table. Also, please fill out the program planning sheet with exercises and other information for your client to follow based on their exercise goals.

MUSCULAR STRENGTH ASSESSMENT

Name _____ Sex _____ Date _____

Age _____   Height _____ cm   Weight _____ kg = _____ lb

Exercise _____

1RM and Estimated 1RM Protocol

☐ Warm-Up Set 1

Weight _____ lb   _____ repetitions   RPE _____

☐ Warm-Up Set 2

Weight _____ lb   _____ repetitions   RPE _____

☐ Warm-Up Set 3

Weight _____ lb   _____ repetitions   RPE _____

☐ Trial Set 1

Weight _____ lb   _____ repetitions   RPE _____

☐ Trial Set 2

Weight _____ lb   _____ repetitions   RPE _____

☐ Trial Set 3

Weight _____ lb   _____ repetitions   RPE _____

☐ Trial Set 4

Weight _____ lb   _____ repetitions   RPE _____

☐ Trial Set 5

Weight _____ lb   _____ repetitions   RPE _____

1RM _____ lb

Relative Strength (1RM [lb] / Body Mass [lb]) _____

Percentile Rating _____

From J. Janot and N. Beltz, *Laboratory Assessment and Exercise Prescription* (Champaign, IL: Human Kinetics, 2023).

RESISTANCE TRAINING PROGRAM PLANNING SHEET

Complete your planar chart with exercises that focus on all planes of motion for each major joint. Using your planar chart, mix and match exercises to complete the program that follows.

Planar Chart

Exercise	Major joint(s)[a]	Plane(s) of motion[b]	Movement direction(s)[c]	Muscle group(s)
Exercise 1				
Exercise 2				
Exercise 3				
Exercise 4				
Exercise 5				
Exercise 6				
Exercise 7				
Exercise 8				
Exercise 9				
Exercise 10				
Exercise 11				
Exercise 12				
Exercise 13				
Exercise 14				
Exercise 15				
Exercise 16				
Exercise 17				
Exercise 18				
Exercise 19				
Exercise 20				

[a]Joints: shoulder, elbow and wrist, hip, knee, ankle, trunk

[b]Planes: sagittal, frontal, transverse

[c]Movement directions: flexion and extension, abduction and adduction, internal and external rotation, horizontal abduction and adduction

> continued

Resistance Training Program Planning Sheet   *> continued*

Monday: Full Body

Exercise	Sets	Reps	Weight (lb)	Rest period	RPE
Exercise 1					
Exercise 2					
Exercise 3					
Exercise 4					
Exercise 5					
Exercise 6					
Exercise 7					
Exercise 8					

Wednesday: Full Body

Exercise	Sets	Reps	Weight (lb)	Rest period	RPE
Exercise 1					
Exercise 2					
Exercise 3					
Exercise 4					
Exercise 5					
Exercise 6					
Exercise 7					
Exercise 8					

Friday: Full Body

Exercise	Sets	Reps	Weight (lb)	Rest period	RPE
Exercise 1					
Exercise 2					
Exercise 3					
Exercise 4					
Exercise 5					
Exercise 6					
Exercise 7					
Exercise 8					

From J. Janot and N. Beltz, *Laboratory Assessment and Exercise Prescription* (Champaign, IL: Human Kinetics, 2023).

Assessment of Clinical Variables: Pulmonary Function Testing

Purpose of the Lab

A pulmonary function test (PFT) is used to assess the presence and progression of pulmonary diseases. Completing a battery of pulmonary function testing highlights the overall functional ability of the respiratory system but does not have the clinical sensitivity to diagnose specific diseases. The respiratory patterns can assist practitioners in understanding the severity of underlying disease, efficacy of current treatment, and appropriateness of alternative treatment options (Gold and Koth 2016).

There are many different types of pulmonary disorders, but the two main categories that we will examine in this chapter are obstructive and restrictive lung disease. Obstructive diseases such as chronic obstructive pulmonary disease (COPD), emphysema, and asthma comprise ~80% of the functional lung diseases. The most common of these obstructive lung diseases is COPD. COPD is mostly caused by prolonged exposure to noxious air par-

ticles such as cigarette smoke or air pollutants. The chronic exposure leads to an enhanced inflammation response in the small airways, mucous hypersecretion, and lung tissue destruction (MacNee 2006). Obstructive lung disease causes resistance to airflow due to an obstruction at any level between the trachea and the terminal bronchioles. Regarding the PFT during this lab activity, we will see that obstructive disease displays a functional pattern of inability to empty the lungs adequately but the patient will often have normal total lung volumes. In contrast, restrictive diseases comprise ~20% of pulmonary disease cases and reduce the overall distensibility or compliance of the lungs, impairing overall lung volumes. This results in an inability to adequately fill the lungs during a PFT, but an individual may have preserved flow rates (Martinez-Pitre, Sabbula, and Cascella 2021). Common restrictive lung diseases include pulmonary fibrosis, pneumonia, and obesity. In pulmonary fibrosis, inflammation within the tissues involved in gas transfer causes deposition of collagen in the tissue layers. Over time, this causes a thickening in the space used during gas exchange and a reduction in lung compliance (Martinez-Pitre, Sabbula, and Cascella 2021).

We will examine obstructive lung disease, restrictive lung disease, and a mixture of these diseases in the current lab by directly measuring the ventilation ($\dot{V}_E$) of our clients. Ventilation is a function of the amount of air within each breath, or tidal volume (V_t), and the number of breaths taken per minute, or respiratory rate (RR). Therefore, $\dot{V}_E$ (L·min^{-1}) = V_t (L·breath^{-1}) $\times$ RR (breaths·min^{-1}). We measure V_t and other functional lung volumes in a testing procedure known as forced spirometry, plotting maximal inhalation and exhalation over time. According to Gold and Koth (2016), indications for performing spirometry testing include evaluating occupational hazards in workers with a higher incidence of lung disease, identifying COPD risk in smokers, quantifying the level of disability within individuals with lung conditions, evaluating the efficacy of treatment in patients with asthma, and tracking the progress of various pulmonary illnesses. There are different functional volumes in the lungs that can be measured during a PFT (figure 12.1).

The specific volumes that we will collect during this lab are forced vital capacity (FVC), forced expiratory volume (FEV), residual lung volume (RV), and maximal voluntary ventilation (MVV). FVC is the largest volume of air measured after forced expiration following full inspiration. FEV is the amount of air expired after a set amount of time,

most commonly within the first second (FEV$_{1.0}$). RV is the amount of air remaining in the lungs after maximal exhalation. The purpose of RV is to keep the alveoli inflated, providing the means for continuous gas exchange even after maximal exhalation. We will calculate RV in this lab for its use in the hydrostatic weighing equation (lab 4). MVV is the volume of air that an individual can breathe with maximal effort over a 10 to 15 s period. Fitness status will be evident during MVV testing because it is more of a performance test to evaluate exercise tolerance and respiratory muscle fitness. The MVV is decreased in clients with obstructive disease and may indicate ventilatory defects, neuromuscular deficits, musculoskeletal disease of the chest wall, or deconditioning due to chronic illness (Gold and Koth 2016). MVV also has value in a preoperative evaluation setting due to the importance of extrapulmonary factors during surgical recovery (Gaensler and Wright 1966). Along with disease status, the main factors that affect these lung volumes are age, height, sex, and ethnicity. Interestingly, fitness status does not have a measurable impact on lung volumes (FVC), but highly trained individuals may be able to generate greater flow rates (FEV and MVV) (Gold and Koth 2016; Haff and Dumke 2019).

The purpose of this lab is to first calculate the predicted lung volumes and lower limit of normal (LLN) values for your client using their age, height, and race. Then, you will conduct a PFT using spirometry (figure 12.2) on your client to evaluate their risk for obstructive, restrictive, or mixed pulmonary disease patterns. Residual lung volume measurement using open circuit nitrogen (N_2) washout will be described in detail within this lab, and the results can be used in this lab and in lab 4.

Necessary Equipment or Materials

- Stadiometer
- Nose clips
- Pneumotach spirometer
- Open circuit N_2 washout system
- Pulmonary function testing data sheet

Calibration of Equipment

Calibration of the spirometry and N_2 washout systems should be performed in accordance with the manufacturer guidelines. Generally, these sys-

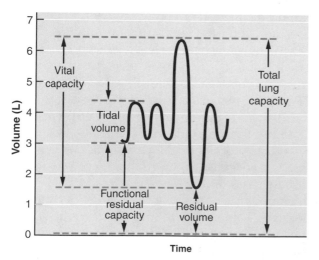

FIGURE 12.1 Spirometry tracing of lung volumes and capacities.

Reprinted by permission from M.H. Malek, "Cardiorespiratory System and Gas Exchange," in *Essentials of Personal Training*, 2nd ed., edited for the National Strength and Conditioning Association by J.W. Coburn and M.H. Malek (Champaign, IL: Human Kinetics, 2012), 25.

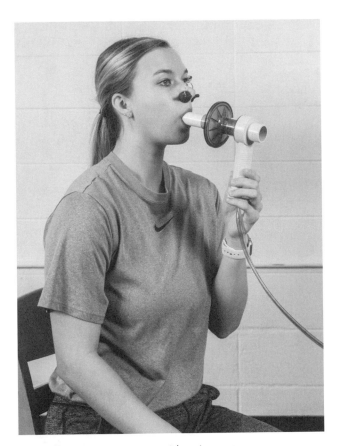

FIGURE 12.2   Participant with spirometer.

tems need to warm up for at least 30 min prior to calibration. Spirometry systems may require flow calibration of the pneumotach, while N_2 washout systems may require gas calibration.

Procedures

Pretest Guidelines

Although PFT is generally safe, there are some contraindications involving specific conditions (Ranu, Wilde, and Madden 2011). These include the following:

- Unstable angina
- Myocardial infarction within the past month
- Recent ophthalmic surgery
- Recent thoracoabdominal surgery
- Thoracic or abdominal aneurysm
- Current pneumothorax

Additionally, it is best practice not to perform PFT on individuals who are contagious with upper or lower respiratory infections, for the purpose of minimizing disease transmission.

Testing

Estimation Equations

1. Gather and record the client's height, age, and race on the pulmonary function testing data sheet.
2. Using the equations that follow, calculate your client's predicted FVC, $FEV_{1.0}$, and RV (Coates et al. 2016; Stocks and Quanjer 1995). Record the predicted lung volume on the pulmonary function testing data sheet.

   Predicted lung volumes for men (≥18 years old):

   $$FVC = -0.183971 + 0.000356\,(Age) - 0.000288\,(Age)^2 + 0.00018356\,(Height)^2$$

   $$FEV_{1.0} = 0.992494 - 0.027146\,(Age) + 0.00013066\,(Height)^2$$

   $$RV = 0.0131\,(Height) + 0.022\,(Age) - 1.232$$

   Predicted lung volumes for women (≥18 years old):

   $$FVC = -0.103966 + 0.013989\,(Age) - 0.000369\,(Age)^2 + 0.00014611\,(Height)^2$$

   $$FEV_{1.0} = 0.676960 - 0.012968\,(Age) - 0.000105\,(Age)^2 + 0.00011234\,(Height)^2$$

   $$RV = 0.01812\,(Height) + 0.016\,(Age) - 2.003$$

3. Using the equations that follow, calculate your client's LLN for FVC, $FEV_{1.0}$, and $FEV_{1.0}/FVC$ (Coates et al. 2016).

   Lower limit of normal for men (≥18 years old):

   $$FVC = 0.223088 + 0.000962\,(Age) - 0.000312\,(Age)^2 + 0.00014013\,(Height)^2$$

   $$FEV_{1.0} = 1.011663 - 0.027985\,(Age) + 0.00010484\,(Height)^2$$

   $$FEV_{1.0}/FVC = 73.995473 - 0.124630\,(Age)$$

   Lower limit of normal for women (≥18 years old):

   $$FVC = -0.640178 + 0.015922\,(Age) - 0.000411\,(Age)^2 + 0.00013922\,(Height)^2$$

   $$FEV_{1.0} = 0.375791 - 0.006589\,(Age) - 0.000180\,(Age)^2 + 0.00009577\,(Height)^2$$

   $$FEV_{1.0}/FVC = 77.743651 - 0.124630\,(Age)$$

4. If needed, apply the most appropriate race-specific correction factor for predicted and LLN FVC and $FEV_{1.0}$ values (Coates et al. 2016) (table 12.1). Reliable correction

factors are not available for $FEV_{1.0}/FVC$. Record the values on the pulmonary function testing data sheet.

Spirometry

Each PFT system may have slightly different methods for collecting spirometry data. Consult the user manual for your system to ensure proper execution of the test. The general guidelines for performing spirometry testing are listed here.

1. If required, install a filter into the pneumotach device on the PFT system.
2. The client should be seated comfortably because they may experience dizziness during testing.
3. Instruct the client to place the nose clips and explain the following testing sequence.
   a. Place your lips around the mouthpiece or end of the pneumotach filter to form a tight seal.
   b. Take two or three normal breaths.
   c. Perform a maximal inspiration, sitting up straight and filling the lungs.
   d. Once the lungs are filled, expire with maximal force. You may want to bend over to engage all the respiratory muscles during expiration.
   e. Continuously exhale for at least 6 to 8 s.
   f. Take a maximal effort inspiration to finish the test.
4. Repeat this testing sequence until two trials agree within 10% of the highest value.
5. Record $FEV_{1.0}$, FVC, and $FEV_{1.0}/FVC$ data for all trials on the pulmonary function

testing data sheet. Circle the best trial for each of the values.

Maximal Voluntary Ventilation

1. Calculate predicted MVV using the following equation (Campbell 1982):

$$MVV \ (L \cdot min^{-1}) = FEV_{1.0} \times 40$$

2. If required, install a filter into the pneumotach device on the PFT system.
3. The client should be seated comfortably because they may experience dizziness during testing.
4. Instruct the client to place the nose clips and explain the following testing sequence.
   a. Place your lips around the mouthpiece or end of the pneumotach filter to form a tight seal.
   b. Take two or three normal breaths.
   c. Breathe as rapidly and deeply as possible for the duration of the test, commonly 15 s in duration.
5. Repeat this testing sequence until two trials agree within 10% of the highest value.
6. Record all trials on the pulmonary function testing data sheet. Circle the best trial.

Residual Lung Volume

To estimate RV, most RV systems use either open circuit N_2 washout—sometimes called *oxygen dilution*—or a closed circuit helium washout method. Both methods estimate RV by using physiologically inert gases that have poor solubility in the

TABLE 12.1   Race-Specific Correction Factors for FVC and $FEV_{1.0}$ Values

Race	FVC correction factor	$FEV_{1.0}$ correction factor
Aboriginal	0.9970	1.0072
Chinese	0.9474	0.9430
South Asian	0.8889	0.9003
Black	0.8582	0.8768
Latin	1.0112	1.0213
Filipino	0.9054	0.9189
Southeast Asian	0.9375	0.9365

Data from Coates et al. (2016) and Stocks and Quanjer (1995).

lungs and alveolar capillaries. Initial concentrations of N_2 are calculated by the RV system by entering atmospheric information and assuming specific gas fractions and N_2 elimination rates. The concentrations of exhaled gases from the client are continuously measured after the client maximally expires and then breathes 100% oxygen for a specific duration. The amount of N_2 "washed out" during this process allows for the calculation of RV. A general description of the open circuit N_2 washout technique follows, but consult the manual on your system for the RV testing procedure.

1. Each trial will require a three-way mouthpiece attached to a 5 L bag of 100% oxygen.
2. The client should remain seated during the entire testing process.
3. Instruct the client to place the nose clips and explain the following testing sequence.
    a. Place your lips around the mouthpiece or end of the three-way valve to form a tight seal. (The technician should ensure the three-way valve is in the position open to room air breathing. There should not be any visible fluctuations in the oxygen bag during this part of the test.)
    b. Take two or three normal breaths.
    c. Perform a maximal inspiration, sitting up straight and filling the lungs.
    d. Once the lungs are filled, expire with maximal force. You may want to bend over to engage all the respiratory muscles during expiration.
    e. Continuously exhale.
    f. Once you are confident that all air has been expired, give a thumbs up to the technician.
    g. Once the technician turns the three-way valve to the 100% oxygen bag breathing position, they will instruct you to take a deep inhale and continue breathing deeply and slowly for eight breaths. This breathing sequence will take ~30 s.
4. Repeat this testing sequence until two trials agree within 10% of the highest value.
5. Record all trials on the pulmonary function testing data sheet.
6. Use the average of the two trials within 10% as the measured RV.

Common Errors

- The client performs a submaximal inhalation prior to maximal exhalation. Ensure that the client completely fills their lungs prior to maximal exhalation.
- After maximal inhalation, the client slightly hesitates in exhalation prior to maximal force exhalation. Ensure that the client performs a maximal exhalation immediately when instructed.
- The client performs a submaximal exhalation. Encourage the client to blast the air out as hard as they can.
- The client may cough during the exhalation phase.
- The client may not exhale for the entire 6 to 8 s duration of the test. Encourage them to keep exhaling until you have determined that 6 to 8 s has been reached.
- During MVV testing, the client may give a submaximal effort. Encourage them to breathe as deep, hard, and fast as they can until the 15 s test is completed.
- During RV testing, the client gives a submaximal effort when emptying their lungs. Encourage the use of all respiratory muscles to exhale as much as possible. Tell them to continue pushing air out until they are certain that all air has been removed from their lungs.

Normative Data and Interpretation

Figure 12.3 shows three distinct flow-volume loop patterns. You can immediately see the difference in flow-volume loop patterns throughout maximal expiration and inspiration. The normal spirometry pattern (figure 12.3a) displays a sharp increase in expiratory flow, peaking at ~8.0 $L \cdot s^{-1}$ and then slowing in velocity until reaching the FVC of ~6.0 L. Compare the normal flow-volume loop to the restrictive lung disease pattern (figure 12.3b) and you will see a tall and narrow flow-volume curve with expiratory flow velocity peaking at ~7.75 $L \cdot s^{-1}$ but a reduced FVC of ~2.0 L. This demonstrates a limitation in the ability to fill the lungs. Looking at the obstructive lung disease pattern (figure 12.3c),

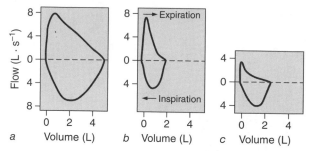

FIGURE 12.3 Examples of three typical pulmonary function testing flow patterns. Flow-volume spirometry in normal (*a*), restrictive (*b*), and obstructive (*c*) patterns.

Reprinted by permission from W.M. Gold and L.L. Koth, "Pulmonary Function Testing," in V.C. Broaddus, R.J. Mason, J.D. Ernst, et al., *Murray and Nadel's Textbook of Respiratory Medicine*, 6th ed. (Philadelphia: W.B. Saunders, 2016), 407-435e18.

the biggest difference is the extreme reduction in expiratory flow velocity. It is not uncommon to see a reduction in FVC as well, but the peak expiratory flow velocity is ~3.75 L·s^{-1}. The flow-volume loop pattern in obstructive lung disease looks similar in morphology to a nondiseased pattern, but it has much smaller maximal flow velocity and total volume. This shows a limitation in the ability to empty the lungs due to the loss in elastic recoil within the airways. The main purpose of performing a PFT in an exercise physiology laboratory is to gather data that may be suggestive of an underlying pulmonary disease or to track disease progression in individuals with known pulmonary disease. Therefore, the diagnosis and medical interpretation of pulmonary diseases are at the discretion of a healthcare provider rather than the exercise physiologist.

Johnson and Theurer (2014) developed an algorithm to simplify the interpretation of PFTs. We will use a modified version of this algorithm given the availability of materials and purpose of this lab. The interpretation involves answering three questions.

Q1: Is the FEV$_{1.0}$/FVC low?

The first step in PFT interpretation is to determine whether the FEV$_{1.0}$/FVC is low. The recommended way to determine this is to compare the client's measured value to the predicted LLN FEV$_{1.0}$/FVC. Locate the predicted LLN FEV$_{1.0}$/FVC and measured FEV$_{1.0}$/FVC on your pulmonary function testing data sheet.

- Q1 Yes: This is suggestive of either an obstructive pulmonary disease or a mixed

pulmonary disease pattern. Proceed to Q2 for further interpretation.

- Q1 No: This is suggestive of either a restrictive pulmonary disease pattern or normal PFT. Proceed to Q2 for further interpretation.

Q2: Is the FVC low?

A low FVC can indicate the presence of restrictive pulmonary disease. The recommended way to determine whether FVC is low is to compare the client's measured value to the predicted LLN FVC. Locate the predicted LLN FVC and measured FVC on your pulmonary function testing data sheet.

- Q1 Yes, Q2 Yes: This indicates a mixed pulmonary disease pattern, suggesting presence of both obstructive and restrictive disease patterns. Proceed to Q3.

- Q1 Yes, Q2 No: This suggests presence of obstructive pulmonary disease. Proceed to Q3 for further interpretation.

- Q1 No, Q2 Yes: This suggests presence of restrictive pulmonary disease. Proceed to Q3 for further interpretation.

- Q1 No, Q2 No: This suggests an absence of pulmonary disease. The interpretation is concluded to be "normal."

Q3: What is the severity of the pulmonary disease?

Calculate the percentage of predicted FEV$_{1.0}$ for your client by dividing your client's measured FEV$_{1.0}$ by their predicted FEV$_{1.0}$ and multiplying by 100.

$$\text{Percentage of Predicted FEV}_{1.0} = (\text{Measured FEV}_{1.0} / \text{Predicted FEV}_{1.0}) \times 100$$

Locate predicted and measured FEV$_{1.0}$ values on your pulmonary function testing data sheet. Compare your calculated percentage of predicted FEV$_{1.0}$ to the values that follow on the severity of pulmonary disease based on percentage of predicted FEV$_{1.0}$ to answer Q3 (Pellegrino et al. 2005):

- >70%: Mild
- 60%-69%: Moderate
- 50%-59%: Moderately severe
- 35%-49%: Severe
- <35%: Very severe

Case Study for Pulmonary Function Testing

A client in your community-based exercise program completes a routine PFT every spring. This client has a family history of pulmonary disease. Testing is performed at your facility, and the report is provided to their primary care healthcare provider. The client's demographic information and the results of their PFT are listed here:

Age: 58
Height: 5 ft 8 in. (173.9 cm)
Sex: Male
Race or Ethnicity: Chinese

Prediction Calculations

$$FVC = -0.183971 + 0.000356 (58) - 0.000288 (58)^2 + 0.00018356 (173.9)^2 = 4.42 \text{ L}$$

$$\text{Adjusted for race} = 4.42 \text{ L} \times 0.9474 = 4.19 \text{ L}$$

$$FEV_{1.0} = 0.992494 - 0.027146 (58) + 0.00013066 (173.9)^2 = 3.37 \text{ L}$$

$$\text{Adjusted for race} = 3.37 \text{ L} \times 0.943 = 3.18 \text{ L}$$

$$MVV = 3.18 \text{ L} \times 40 = 127.2 \text{ (L·min}^{-1})$$

$$RV = 0.0131 (173.9) + 0.022 (58) - 1.232 = 2.32$$

LLN Calculations

$$FVC = 0.223088 + 0.000962 (58) - 0.000312 (58)^2 + 0.00014013 (173.9)^2 = 3.47 \text{ L}$$

$$\text{Adjusted for race} = 3.47 \text{ L} \times 0.9474 = 3.29 \text{ L}$$

$$FEV_{1.0} = 1.011663 - 0.027985 (58) + 0.00010484 (173.9)^2 = 2.56 \text{ L}$$

$$\text{Adjusted for race} = 2.56 \text{ L} \times 0.943 = 2.41 \text{ L}$$

$$FEV_{1.0}/FVC = 73.995473 - 0.124630 (58) = 66.8\%$$

$$\% \text{ of predicted} = (\text{best} \div \text{predicted}) \times 100$$

PULMONARY FUNCTION TESTING RESULTS

	Trial 1	Trial 2	Best	Predicted	LLN	% of predicted
FVC (L)	3.62	3.51	3.62	4.19	3.29	86
FEV$_{1.0}$ (L)	1.92	1.89	1.92	3.18	2.41	60
FEV$_{1.0}$/FVC (%)	53.0	53.8	53.8	75.9	66.8	71
MVV (L·min^{-1})	95.3	97.9	97.9	127.2	–	77
RV (L)	2.68	2.84	2.76*	2.32	–	119

*Average of the two trials.

Note: FVC = forced vital capacity; FEV$_{1.0}$ = forced expiratory volume in the first second; MVV = maximal voluntary ventilation; RV = residual lung volume.

Q1: Is the FEV$_{1.0}$/FVC low?
Yes, the client's best FEV$_{1.0}$/FVC trial of 53.8% is below the LLN of 66.8%.

Q2: Is the FVC low?
No, the client's best FVC of 3.62 L is above the LLN of 3.29 L.

Q3: What is the severity of the pulmonary disease?
The percent of predicted for FEV$_{1.0}$ was calculated at 60%.

Q1 was answered as a "Yes" and Q2 was answered as a "No." This means that the PFT indicated a pattern of obstructive pulmonary disease. The 60% of predicted FEV$_{1.0}$ suggests a moderate disease severity. We can also see that MVV was 77% of predicted and that "trapping of air" was evident by a higher than predicted RV; both findings may be used by a healthcare provider to further confirm the presence of an obstructive pulmonary disease.

PULMONARY FUNCTION TESTING

Name _____ Sex _____ Date _____

Age _____ Height _____cm Race _____

Predicted Lung Volume Calculations

FVC = _____

$FEV_{1.0}$ = _____

RV = _____

MVV = _____

Lower Limit of Normal Calculations

FVC = _____

$FEV_{1.0}$ = _____

$FEV_{1.0}/FVC$ = _____

	Trial 1	Trial 2	Trial 3		Best	Predicted	LLN	% of predicted
FVC (L)								
$FEV_{1.0}$ (L)								
$FEV_{1.0}/FVC$ (%)								
MVV (L·min^{-1})							–	
RV (L)*							–	

*Use average of two trials within 10% for RV.

Q1: Is the $FEV_{1.0}/FVC$ low? (circle one)   Y   N

Q2: Is the FVC low? (circle one)   Y   N

Q3: What is the severity of the pulmonary disease? _____

Interpretation:

_____

_____

_____

_____

_____

_____

_____

From J. Janot and N. Beltz, *Laboratory Assessment and Exercise Prescription* (Champaign, IL: Human Kinetics, 2023).

Assessment of Clinical Variables: Basic Electrocardiography

Purpose of the Lab

An electrocardiogram (ECG) is an aggregate recording of the electrical activity occurring in the heart. This technology, first developed and modernized by the Dutch physician and physiologist Willem Einthoven, dates to the early 20th century (AlGhatrif and Lindsay 2012). The ECG can give us a direct assessment of electrical function and an indirect assessment of structural issues of the heart (Goldberger, Goldberger, and Shvilkin 2013). In this instance, function refers to the electrical conductivity of impulses flowing through the heart. From the ECG, we can tell whether electrical conduction is happening normally or abnormally and where exactly in the heart electrical impulses are being generated and conducted from. Structure refers to potential abnormalities of the myocardium itself, such as with a previous myocardial infarction, current myocardial ischemia, or if a heart has experienced hypertrophy or enlargement. Structural issues are depicted on the ECG in a specific way only because cardiac electrical conductance is affected by these abnormalities (Goldberger, Goldberger, and Shvilkin 2013).

The electrocardiograph is the machine that records the ECG, using electrodes placed on the surface of the skin that will sense cardiac electrical activity. The electrodes are connected to wires used to transmit what they are sensing back to the electrocardiograph. The electrocardiograph, using single or pairs of electrodes along with these wires, forms what we call leads. A lead is a view of the cardiac electrical activity from a specific orientation unique to that lead. In this lab, we will be able to look at the electrical activity from 12 different orientations using what we call the 12-lead ECG. The view of the heart also exists in two different planes: frontal and horizontal. This results in a three-dimensional summation of the electrical activity happening during each heartbeat.

The primary purpose of this lab is to introduce you to the procedures involved in the preparation of the 12-lead ECG and provide experiences to practice this skill with your classmates. You will also have an opportunity to record and interpret your own 12-lead ECG. The secondary purpose of this lab is to teach you the underlying principles behind basic cardiac rhythm interpretation and provide you with practice analyzing and identifying rhythms.

Necessary Equipment or Materials

- Non-latex gloves
- Soft scrubber abrasive pads for skin preparation
- Nonpermanent marker to mark electrode sites
- Premoistened alcohol pads (or gauze pads and 70% isopropyl alcohol)
- Electric clippers or disposable dry-shave razor for body hair removal
- ECG electrodes (adult size) and electrode gel (if needed)
- Towel(s)
- ECG monitor, printer, and paper
- Table with cushioned mat (therapy table preferred)
- Resting 12-lead ECG interpretation lab sheet
- ECG guidelines to know lab sheet
- ECG rhythm recognition practice sheet

Note: Women should be instructed to wear sports bras for this lab for easier placement of the precordial or chest electrodes and to lessen the amount of motion artifact (noise) in the precordial leads.

Calibration of Equipment

For this lab, no calibration of equipment is needed. Make sure that all equipment is in good working condition before use. Check that all lead wires are present, make sure that they are not frayed or damaged, and check that all electrode clips are clean and are making good contact with the electrodes. Also, check the electrocardiograph to make sure that there is sufficient printer ink and paper to run your test.

Procedures

The first part of the ECG lab requires students to work in small groups (2-3 students) to prepare and record 12-lead ECGs. Electrode placement site identification takes the most skill and practice; thus, practice identifying these anatomical sites on more than one classmate, but only prepare one 12-lead ECG per person. You will complete three separate ECG recordings in the following order: (1) following 5 min of supine rest, (2) 20 s following a supine to standing change of position, and (3) immediately following 10 s of hyperventilation while seated. These will be printed and analyzed as part of the tasks for this lab.

The second part of the lab involves an overview and practice of ECG rhythm interpretation. In this portion of the lab, you will fill out the ECG guidelines to know lab sheet as you read and learn about the various characteristics of the electrical cardiac cycle. Following this and a review of basic information regarding various cardiac rhythms, you will practice interpreting a set of rhythm strips on the ECG rhythm recognition practice sheet. If it is helpful, work with your group members and solve these rhythms together. Once finished, you can check your answers against the key provided to you.

Procedures for Skin Preparation

This section details the steps for preparing the client's skin for electrode placement. Before beginning, make sure that all supplies are accessible to you and ready to use, and the electrocardiograph is powered up and ready for recording. Prepping the 12-lead ECG takes time, so it is vital that you are prepared and working efficiently. The procedures for skin preparation are as follows:

1. Organize your prepping supplies and have 10 electrodes ready for placement. Make sure that the electrode bag is zipped tight to prevent unused electrodes from drying out.

2. Review with your client the steps involved with skin preparation and electrode placement. Some clients can be anxious about the use of skin scrubbers and alcohol that might irritate their skin, and the potential use of razors to remove hair. Inform your client that the alcohol and electrode gel can be cold at first on their skin. It is always best to explain procedures and address questions that they may have up front.

3. Put on a pair of tight-fitting non-latex gloves and have the client lay on a therapy table or a regular table with a mat under them for comfort.

4. Ask your client to lift their shirt up toward their shoulders to provide access for prepping the electrode sites. For clients who are women, place a towel over their sports bra or breast area for the purpose of being discreet. Remember to maintain professional behavior and relaxed confidence when working with clients and be sure to tell them what you are going to do before performing each step in this process. Additionally, do not lay any ECG preparation supplies anywhere on your client; always keep them off to the side.

5. Locate the sites for electrode placement by following the directions in the Procedures for Electrode Placement section that follows.

6. If sites need to be cleared of hair, use a disposable dry-shave razor or clippers. Make long, sweeping strokes when using the razor or clippers to remove the body hair. Once the hair is removed, mark the specific electrode placement site with a nonpermanent marker.

7. Take a small abrasive pad and scratch the skin at all marked sites. This is done to increase surface area contact for electrodes and reduce resistance to sensing cardiac electrical activity. Use small, sweeping strokes to rough up the skin. This should be done from top to bottom (not back and forth) and repeated, like sanding wood, for instance. Use light pressure on the skin and repeat until the skin looks dry and chalky. On lighter skin, a red or pink appearance to the skin may occur.

8. After abrading the skin, use the alcohol pads to remove oils, dirt, dead skin cells, moisturizer, and so forth from the skin. This will also enhance cardiac electrical activity sensing by the electrode. While applying moderate pressure, rub back and forth or in a circular motion on the skin approximately five to six times. Do not touch the area once skin cleaning has been done.

9. The electrodes can now be placed. Center each electrode *exactly* on the spot that you have marked on the skin. Once this is done, you can attach the lead wires to each electrode.

10. After recording your ECGs, assist your client as needed to remove the electrodes from their skin. These are usually much easier to remove after an exercise test versus resting due to sweat loosening up the adhesive on the electrode. In either case, take care when removing the electrode. A good technique to use is to gently pull the client's skin in the opposite direction that you are going to remove the electrode. Fold the electrode over itself slowly as you lift it up off the skin. Roughly removing the electrode and pulling skin with it can lead to skin irritation and, at worst, tearing of the skin in some individuals, such as older adults or those with conditions affecting the skin.

Procedures for Electrode Placement

This section details the steps for electrode placement. Before beginning, make sure that the 10 electrodes are ready to use and in good condition. Damaged or dried out electrodes should be discarded. Check the consistency of the gel on the bottom side of the electrode before taking the plastic off. Gel should be present and the electrode moist. The procedures for electrode placement are as follows.

Limb lead electrode placement. The right arm (RA) and left arm (LA) electrodes are placed approximately 2 cm inferior to the clavicle and away from the left and right deltoids (Francis 2016). The right leg (RL) and left leg (LL) electrodes are placed on the abdomen halfway between the costal (ribs) margin and iliac crest along the anterior axillary line. The placement of these four electrodes is referred to as the modified Mason-Likar configuration. This is done both to minimize motion artifact during exercise and to make exercise ECG assessment possible (ACSM 2021; Mason and Likar 1966). The downside to the modified Mason-Likar system is that the limb lead placement is different from the standard configuration used in diagnostic testing. Any changes from the standard configuration can affect the diagnostic value of the ECG (Francis 2016). Changes to wave forms on the ECG have been reported when comparing the Mason-Likar to standard ECG configurations, which could lead to misdiagnosis of potential conditions (ACSM 2021; Francis 2016). Therefore, it is recommended that if the Mason-Likar system is used, it should be noted on the ECG.

Precordial or chest lead electrode placement (Goldberger, Goldberger, and Shvilkin 2013). For the precordial leads (figure 13.1), we start first with V_1 and V_2. These two electrodes are placed on the right (V_1) and left (V_2) side of the client's sternum in the fourth intercostal space. Anatomically, right and left is always oriented specific to the patient and not the technician. To find the fourth intercostal space, first move your fingers down on the sternum from the sternal notch and palpate for the sternal angle (or angle of Louis), which is a ridge formed by the manubrium joining with the sternum. The second rib is on either side of this ridge. Palpate over to find the second rib, move your fingers below it to find the second intercostal space, then move your fingers down to the third rib and so on until you find the fourth intercostal space. The

next anatomical site is for V_4. Find the midpoint of the left side clavicle and draw an imaginary line down to the fifth intercostal space and place V_4 here. The V_3 electrode goes between V_2 and V_4. The last two sites are for V_5 and V_6. The electrode for V_5 is placed directly in line with V_4 at the left anterior axillary line, and V_6 is placed directly in line with V_5 at the left midaxillary line.

In general, do not place any electrodes on bone, active muscle (e.g., deltoids), or breast tissue. Some electrodes may need to be moved slightly from the specific site to stay away from these structures. If breast tissue is impeding your access to a particular electrode site, likely V_4 or V_5, ask the client to move their breast so you can place the electrode as close as you can to the anatomical site. To be discreet, it is very important that you ask the client to do this themselves. To view the final configuration of electrode placement, see figure 13.2.

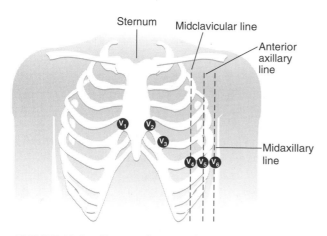

FIGURE 13.1   Client with precordial or chest lead ECG.

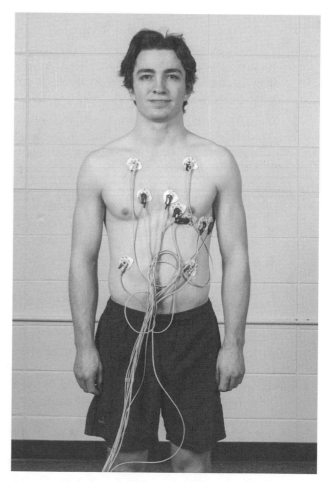

FIGURE 13.2   Client with full 12-lead ECG preparation.

ECG Data Collection Procedures

To complete the first part of this lab, you will record a 12-lead ECG during three different conditions. The first ECG will be recorded with the client in a supine position and at rest. After prepping and connecting the electrodes to the electrocardiograph, have your client rest comfortably for 5 min with arms at their sides and legs uncrossed. View the monitor on the electrocardiograph to ensure that the recording is clear and free from interference and movement artifact. Once the 5 min rest period is up, print the ECG and label it "resting."

After the resting ECG is printed, prepare your client to move to a standing position. Following 20 s of standing with minimal movement to prevent artifact, print another 12-lead ECG and label "standing." Once this is done, ask your client to be seated. For this last recording, you will instruct your client to hyperventilate for 10 s and print an ECG immediately upon finishing the breathing maneuver. This ECG should be labeled "hyperventilation." The client should breathe both deeply and rapidly during this task. There may be some artifact with this recording and it is a real test of your ECG preparation skills, meaning that less artifact = better prep! Once all ECGs are labeled and recorded, compare the three and note any changes to the overall ECG such as heart rate (HR) (e.g., increase or decrease), rhythm (e.g., regular or irregular), and the specific characteristics of the cardiac cycle (e.g., T wave changes, interval changes, etc.) across the different conditions. It is typical to see changes to HR and wave forms across these conditions.

Basic Steps for 12-Lead ECG Interpretation

Let's discuss the basic interpretation of your resting ECG. Please fill out the resting 12-lead ECG interpretation lab sheet with answers for each item. The normative value for comparison column can be filled in with answers from the lab reading and provide you with a comparison for your findings. Much more can be gleaned from the 12-lead ECG, but that is beyond the scope of this lab. If you are interested in learning more about 12-lead ECG interpretation, consulting other texts for more information is rec-

ommended (Dunbar and Saul 2009; Goldberger, Goldberger, and Shvilkin 2013; Thaler 2018).

The 12-lead ECG contains three standard leads (I, II, III), three augmented leads (aVR, aVL, aVF), and six precordial leads (V_1–V_6). Remember, these leads each provide a different view of the cardiac electrical activity. The limb leads, created using the RA, LA, and LL electrodes, view the cardiac electrical activity in the frontal plane, and the precordial leads view it in the horizontal plane. However, you may notice some slight similarities in the way cardiac cycles look among certain leads. That is because leads can be grouped by the areas of the heart that they view. For instance, II, III, and aVF are collectively referred to as the "inferior leads" because they overlook the inferior portion of the heart. Other lead groupings include V_1 through V_4 (septal/anterior leads) and I, aVL, V_5, and V_6 (lateral leads) (Dunbar and Saul 2009). Lead aVR will normally be seen as "negative" (all waves are upside down) when the limb lead wires are connected correctly, so check to make sure that is the case. Also, the R waves in the chest leads should get steadily taller or more "positive" in amplitude from V_1 (small R wave) to V_6 (large R wave). In a normal ECG, this is another clue that you placed the chest electrodes in the correct position. Lastly, the terms positive and negative are in reference to the positioning of a wave in orientation to the baseline or isoelectric line. This line runs through all cardiac cycles present on the ECG. If any wave is deflected above the baseline, then that wave is described as positive; thus, the opposite is true for waves that are negative.

The ECG graph paper is organized into small squares that are 1 mm in width and height, and large boxes that are 5 mm wide and tall. On the vertical axis, we can use these squares to measure the amplitude (or voltage) of the ECG wave forms. On the horizontal axis, we can measure distance in mm or time in seconds. A 1 mm square is equal to 0.04 s in time. A large box (5 mm wide) is equal to 0.2 s in time. These numbers are dictated by the standard speed (25 mm·s^{-1}) at which the paper moves while printing an ECG. See figure 13.3 for a visual of these measurements.

After a general viewing of the 12-lead ECG, calibration of the ECG should be checked. The electrocardiograph produces an upright rectangular box on either the left or right side of the ECG paper for each line that is printed (see figure 13.4 for

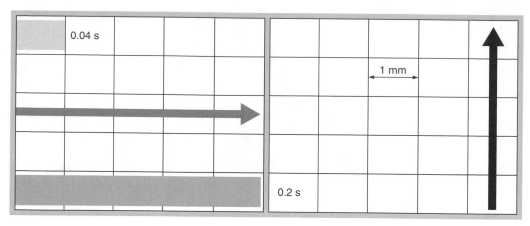

FIGURE 13.3   ECG paper.

an example). This box should be 10 mm in height (equal to 1 mV). This is called standard calibration. This calibration box can be made smaller (5 mm or one-half standard) or taller (20 mm or two times standard) for various reasons if needed. In standard calibration, the amplitude values for wave forms do not need to be adjusted after measurement.

To complete the rest of the assignment, you will first have to work through the second part of the lab that addresses rhythm interpretation. After you finish your resting 12-lead interpretation, review your answers with your group members and seek feedback on your final evaluation. Finally, hand in your resting 12-lead ECG along with the lab sheet to your lab instructor for review.

Common Errors and Assumptions

The main source of error with client preparation and electrode placement lies with the technician. The first step to ensuring a good and accurate representation of the client's ECG tracing is skin preparation. If the steps involved in this task are not followed and performed correctly, the amount of interference and artifact on the ECG will affect the quality and clarity of the recording. If there is too much artifact, it will be difficult to accurately interpret the ECG either on the screen or on paper. A repreparation of the client will need to be done to address any artifact issues. Time and material cost is an issue in this case; thus, prepare the client correctly the first time.

Electrode placement and correct lead wire connection is the next issue affecting the accuracy of the ECG recording. Anatomical landmarks (much like for skinfold measurements) are used to standardize the placement of electrodes in order to obtain the correct representation of the individual client's cardiac electrical activity and function. The ECG guidelines presented in the following section are largely based on correct electrode placement. If electrodes are placed in the wrong location, this can ultimately affect how the ECG is recorded and interpreted. Additionally, ensure that each lead wire is connected to the correct electrode following initial placement. It is recommended to identify the correct lead wire from the diagram on the ECG wire housing (the place where all wires originate from) and match that to the electrode of interest. The first sign of a mix-up of lead wires and their electrodes will be a very strange looking ECG.

Why is all of this so important to do correctly? In cases when physicians or other healthcare providers are using an ECG to diagnose disease or determine underlying causes of chronic conditions, it is essential to have the most accurate information possible. Thus, it is very important that you practice ECG electrode placement and that care is taken to follow all procedures of preparation and placement to get the best result.

Other minor sources of error include electrocardiograph setup prior to assessment. Make sure that recordings are set to standard gain or calibration of 10 mm/mV and paper speed of 25 mm·s^{-1}, which can be checked either on the machine or on most printed ECGs. Fortunately, these standard settings are the default setup for electrocardiographs. Various errors involved with the actual interpretation of the ECG recording will be discussed as necessary in the following sections.

Normative Data and Interpretation

Basic Electrocardiography Application: Cardiac Rhythm Interpretation

In this section, you will learn the various aspects of the electrical cardiac cycle, the step-by-step approach for interpreting cardiac rhythm, and the characteristics of common or more well-known cardiac rhythms. To be successful at developing the skill of cardiac rhythm interpretation, we must take on the role of the "objective detective." The "objective detective" views problems as a puzzle to be solved and attacks those problems with a deliberate approach. Our approach will start with a question that requires a specific answer that will lead to another question, and so on. As we gather information and answers to each question, we will slowly eliminate potential final answers until we are hopefully left with only one possibility at the end of the process.

For this discussion, the term cardiac rhythm refers to the actual "diagnostic" terms that are used to describe the origin and other characteristics that are specific to that cardiac rhythm. The word "rhythm" will also be used in a different context to describe the overall regularity (or irregularity) of electrical cardiac cycles that are recorded on the ECG. The concept of "rhythm" and the recognition of it (i.e., regular versus irregular) will be a crucial step in determining the overall cardiac rhythm interpretation.

The Electrical Cardiac Cycle

Figure 13.4 illustrates all the events that comprise the electrical cardiac cycle. In this example, the cardiac cycle begins with the depolarization of the atria, followed by the depolarization of the ventricles and ending with the repolarization of the ventricles. At rest, the approximate time it takes to conduct the impulse through the cardiac conduction system and repolarize the myocardium so another impulse can be conducted is between 0.48 and 0.52 s. Table 13.1 provides background information on the specific events of the electrical cardiac cycle. It is helpful to refer to figure 13.4 as you review this table.

Let's take a moment to discuss ECG nomenclature specific to the cardiac cycle. A *wave* is a feature on the ECG that has *amplitude*, either deflected in a positive (upward) or negative (downward) direction in relation to the baseline. A *segment* is a flat line that occurs between waves and usually is equal with the baseline. An *interval* is comprised of both a wave(s) and a segment. Lastly, note that the term cardiac cycle can also be applied to the events that occur during the contraction of the heart (i.e., valves opening, blood flowing through or out of

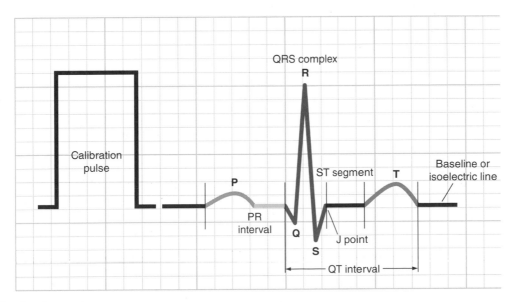

FIGURE 13.4   Cardiac cycle.

TABLE 13.1   The Events of the Electrical Cardiac Cycle, Their Physiological Cause, and Other Details Regarding the Specific Event

ECG event	Physiological cause	Other details
P wave	Atrial depolarization	Because the atria are smaller in terms of muscle mass, the size of the depolarization wave will be smaller compared to the ventricles. P waves are normally small (<2.5 mm high) and narrow (<0.12 s in duration).
PR interval	Atrial, atrioventricular (AV) node, AV bundle (or bundle of His), bundle branches, and Purkinje fiber depolarization	The PR interval is measured from the start of the P wave to the start of the QRS complex. It reflects the time it takes for the impulse to pass from the sinoatrial (SA) node to Purkinje fibers. After moving through the atria, the impulse is slowed slightly in the AV node to allow for ventricular filling. The impulse then travels through the AV bundle, down both left and right bundle branches, and finally to the Purkinje fibers. The impulse passes this relatively long distance in a short period of time. The normal PR interval duration ranges from 0.12 to 0.20 s. In general, anything longer than this would be evidence for a delay in impulse conduction through this part of the pathway.
QRS complex*	See Q, R, and S waves	The impulse moves from the Purkinje fibers to and then through the ventricular myocardium. Since the ventricles together form a large amount of muscle, this impulse needs to travel rather quickly so the ventricles can depolarize and contract together. The normal duration of the QRS complex is 0.06 to 0.10 s. In general, anything longer than this would be indicative of delayed conduction through the ventricles. The QRS is measured from the start of the Q or R wave (if no Q is present) to the end of the S or R wave (if no S is present).
Q wave	Interventricular septum depolarization (Scheidt and Erlebacher 1987)	The negative or downward deflection before the R wave. Normal Q waves are narrow (<0.04 s in duration) and small (measured as <1/3 of its corresponding R wave).
R wave	Early ventricular depolarization (Scheidt and Erlebacher 1987)	The positive or upward deflection in the QRS complex.
S wave	Late ventricular depolarization (Scheidt and Erlebacher 1987)	The negative or downward deflection after the R wave.
ST segment	Slow or early ventricular repolarization	Normally, the ST segment is equal with the baseline. An ST segment that is either above or below the baseline could be a sign of cardiovascular issues.
J point	End of ventricular depolarization and start of repolarization	The J point is the transition point from where the QRS ends and the ST segment begins. It is important to identify where the J point is when evaluating the ST segment.
T wave	Rapid or late ventricular repolarization	T waves are a part of the late ventricular repolarization phase. In general, T waves should be positive or upward in the leads associated with the tallest R waves.
QT interval	Ventricular depolarization and repolarization	The QT interval is measured from the start of the QRS complex to the end of the T wave. This interval comprises the total time it takes for the ventricles to depolarize and fully repolarize, and is influenced by HR. A QT interval of less than 0.44 s is considered normal, with a range of 0.32 to 0.40 s at rest.

*The QRS complex is a collection of waves, thus it is called a complex. However, many different variations of the QRS complex exist. For example, the complex may only contain a Q and R wave, an R and S wave, or possibly just an R wave. Whichever is the case, we will still refer to these as the "QRS complex" to keep it simple.

the heart, etc.). This would be the *mechanical* cardiac cycle. The difference here is that an electrical event precedes any mechanical event. For instance, atrial depolarization occurs first to stimulate the atria to contract and move blood into the ventricles during the "atrial kick." Thus, the electrical event (depolarization) occurs before the mechanical event (atrial contraction).

The Six-Point Rhythm Analysis Checklist

The six-point rhythm analysis checklist is the start-to-finish process that you will employ when interpreting cardiac rhythms. The closer that you follow this checklist, the more likely it will be that no information is missed along the way to the correct interpretation. The steps to our rhythm analysis checklist are as follows:

Step 1: Are there P waves?

The first step is to identify whether P waves are present, which is a simple "yes" or "no" answer. If P waves are present, what do they look like? Small, rounded on top, and narrow at the bottom? Tall and peaked? Notched on top? Positive or negative? As you can tell, P waves can demonstrate a variety of shapes. Establishing their shape plus their uniformity (i.e., do they all look the same or do they differ from cycle to cycle?) matters in identifying cardiac rhythm.

Step 2: What is the rhythm?

Remember, in this context, "rhythm" refers to the regularity or irregularity of the cardiac cycles that are recorded on the ECG. To assess this issue, we will focus on the R wave of each cardiac cycle to determine the ventricular rhythm. For a *regular* ventricular rhythm, the distance between successive R waves (i.e., the R–R interval) is equal and consistent from cycle to cycle (or beat to beat). It is better to measure the R–R interval distance rather than visually inspect. Use a ruler or another measurement device to determine the exact distance between two R waves of your choice. Once you do this, move your measurement device across the ECG rhythm strip measuring the distance between all R waves. If this distance is repeatedly the same, then the ventricular rhythm is regular. If the R–R interval keeps changing from cycle to cycle, then the rhythm is called *irregular*.

Step 3: What is the relationship between the P waves and QRS complexes?

This step involves determining if every P wave that is observed on the ECG is followed by a QRS complex. If there are no P waves present, you can skip this step! If P waves are present, make sure that each one is followed by a QRS and then evaluate the PR interval to see if it falls within normal range. Good practice dictates that we measure the PR interval from a few different cardiac cycles to check for consistency. It is okay to report a range in seconds (e.g., 0.14-0.16 s or 0.18-0.20 s) if the PR interval differs slightly from beat to beat. If the PR intervals differ more than this and are inconsistent from beat to beat, then an arrhythmia may be present.

Step 4: What is the HR?

It is important to establish ventricular rhythm first because rhythm regularity dictates what method to use to determine HR. We have three methods at our disposal to determine HR. Two require a regular ventricular rhythm to work accurately, and the other we use when the rhythm is irregular.

In regular rhythms, we can count the number of large boxes (5 mm width) that occur between one R–R interval and divide 300 by that number. However, you can also do this faster by counting boxes and assigning each one a number until you reach the next R wave. The numbers are assigned to each box in this order: 300, 150, 100, 75, 60, and 50. The number that you stop at in the counting is the HR. The second method involves counting the number of small squares (1 mm width) within one R–R interval and dividing 1,500 by that number. This is helpful when the second R wave in your counting doesn't fall exactly on a solid line of a large box. For instance, if you are counting and you get to the second R wave and it falls within a box halfway between the 75 and 60 number mark, what can you do? You could estimate HR to be halfway between 60 and 75 and say 68 beats·min^{-1}. In many cases, that is probably close enough, or you count the squares and divide 1,500 by that number to get the exact HR. It all depends on how accurate you want to be.

In irregular rhythms, we can't use the first two methods because the R–R interval is constantly changing from cycle to cycle. For this method, our first step is to establish a recording of an ECG rhythm strip that is at least 6 s long. Some electrocardiographs mark 3 s lengths in time at the top or bottom of the ECG as it is printed. If your ECG doesn't make these marks, count out a length of 30 large boxes, which is equal to 6 s of time (30 boxes × 0.20 s per box = 6 s). Once this is established, count the number of R waves that fall within the 6 s strip of time and multiply by 10. This will give you the HR.

Lastly, once HR is established, we need an answer to a secondary question: Is the HR too fast, too slow, or somewhere in the middle? Generally, the terms "too slow" or "too fast" are associated with a HR of <60 beats·min^{-1} and >100 beats·min^{-1}, respectively.

Step 5: What is the shape of the QRS?

For the purposes of this discussion, the shape of the QRS is determined by evaluating the duration of the complex and describing it as either narrow or wide. A narrow QRS falls within the normal limits discussed in table 13.1. A wide QRS would be >0.10 s.

Step 6: What is the cardiac rhythm interpretation?

After gathering all this information, you are ready to make your cardiac rhythm interpretation. Hopefully, following these steps and having a systematic approach to cardiac rhythm and 12-lead ECG

analysis will lead you to the correct answer. It will take a lot of practice to be consistent with this skill, and potentially further study in other courses or practical experience with patients and clients.

Sinus Rhythms

Sinus rhythms originate from the SA node, which is also referred to as the dominant pacemaker in the heart. Pacemaker cells, like in the SA node, AV junction area, and Purkinje fibers, can generate their own impulses without any prior stimulation (Dunbar and Saul 2009). This is called the property of automaticity. The cells in the SA node happen to generate impulses at a regular and faster rate compared to the other pacemaker cell sites, which is the reason why it is the main pacemaker for the heart. Following are the characteristics, including an example rhythm, for each of the sinus rhythms.

Normal Sinus Rhythm

An example of normal sinus rhythm is shown in figure 13.5. The rhythm in normal sinus rhythm is regular and the P waves are small, narrow, and look uniform from cycle to cycle. The QRS complex is narrow and each P wave is followed by a QRS. The HR for normal sinus rhythm is between 60 to 100 beats·min^{-1}.

Sinus Bradycardia

An example of sinus bradycardia is shown in figure 13.6. The rhythm is regular and all other criteria for sinus bradycardia are similar to normal sinus

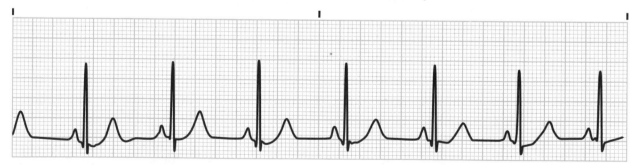

FIGURE 13.5   Normal sinus rhythm.

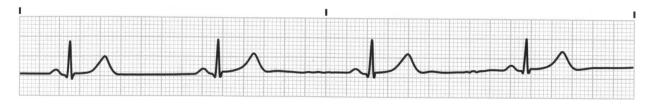

FIGURE 13.6   Sinus bradycardia.

rhythm with the exception of HR. The HR in sinus bradycardia is <60 beats·min⁻¹. Sinus bradycardia can be common in well-trained individuals due to increased parasympathetic nervous system stimulation or vagal tone (Dunbar and Saul 2009). Sinus bradycardia is commonly seen in individuals taking certain antihypertensive medication (see appendix A). Also, sinus bradycardia can be triggered by disease, such as in a myocardial infarction, ischemia of the SA node, and possibly in those experiencing hypotension (i.e., low blood pressure).

Sinus Tachycardia

An example of sinus tachycardia is shown in figure 13.7. The rhythm is regular and all previous criteria are similar apart from HR. The HR in sinus tachycardia is >100 beats·min⁻¹. Sinus tachycardia is commonly seen during exercise or when someone is anxious or nervous. In these cases, sinus tachycardia is due to increased sympathetic nervous system stimulation, also referred to as the "fight-or-flight" system (Scheidt and Erlebacher 1987). Sinus tachycardia can also be caused by taking certain substances like caffeine, nicotine, or other stimulants. Lastly, sinus tachycardia may be seen in someone with heart failure or who is hypoxic due to lack of oxygen.

Sinus Arrhythmia

An example of sinus arrhythmia is shown in figure 13.8. All criteria for normal sinus rhythm are applied to sinus arrhythmia except for rhythm, which is irregular. It is common to see HR <60 beats·min⁻¹ in this rhythm. In this case, irregular is defined by a rhythm where the longest measured R–R interval exceeds the shortest measured R–R interval by 0.16 s (four small squares) or more (Scheidt and Erlebacher 1987). Some criteria put this difference at 0.08 s (two small squares) or more (Dunbar and Saul 2009). Therefore, you will need to pay close attention to the R–R interval and compare the longest and shortest if there is variation from cycle to cycle. Sinus arrhythmia is usually a normal finding and can be observed in a variety of individuals (e.g., well-trained, younger, older). The most common cause of sinus arrhythmia is the respiratory cycle of inspiration and expiration during quiet breathing (Dunbar and Saul 2009; Goldberger, Goldberger, and Shvilkin 2013). As an individual breathes in, HR increases (R–R interval shortens) due to less parasympathetic stimulation; in contrast, HR decreases (R–R interval lengthens) during expiration due to more parasympathetic stimulation.

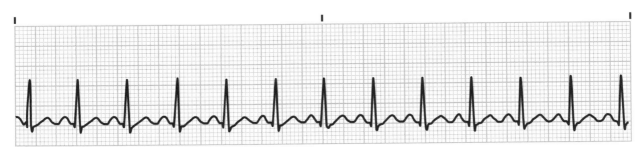

FIGURE 13.7   Sinus tachycardia.

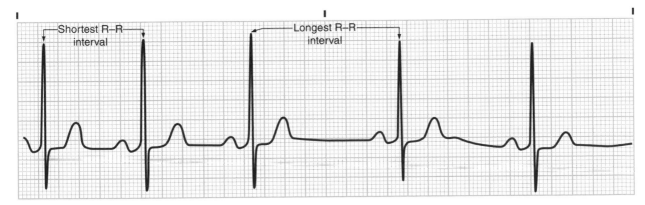

FIGURE 13.8   Sinus arrhythmia.

Atrial Rhythms

Atrial rhythms originate from specialized cells in the atria that can generate impulses. In some cases, these cells can be stimulated through increased sympathetic nervous system activation, stimulant-type drugs and medications, or electrolyte imbalances. In comparison, these rhythms can also be triggered by chronic heart disease linked to ischemia, heart failure, cardiac valve disease, or hypoxic conditions. A full review of the atrial rhythms is beyond the scope of this book, so we will focus on one of the most common atrial rhythms: atrial fibrillation.

An example of atrial fibrillation is shown in figure 13.9. The first feature most people notice is that there are no discernable P waves present in this rhythm. In atrial fibrillation, the atria do not depolarize and contract in coordination as they do in sinus rhythms; instead, they twitch or quiver (fibrillate) in place, which leads to a "scratchy" looking baseline appearance observed in figure 13.9. This baseline appearance is caused by multiple areas sending out impulses in a chaotic fashion within the atria. Thus, no P wave is inscribed on the ECG to signify organized atrial depolarization. This chaotic firing of impulses also leads to another important characteristic of atrial fibrillation: an irregular rhythm. The QRS is usually narrow because once impulses leave the atria, they travel through the AV node and into the myocardium like in the sinus rhythms.

A big challenge in treating atrial fibrillation is the increased risk of blood clot formation in the heart. These clots can break away and get lodged in a blood vessel downstream, such as in the brain, which can lead to a stroke. Thus, atrial fibrillation is treated and managed early and carefully. Often, atrial fibrillation can be controlled by select cardiac medications (see appendix A) or by an internal pacemaker to regulate HR and cardiac rhythm.

Ventricular Rhythms

Ventricular rhythms originate from specialized pacemaker cells within the ventricles, likely from the area of the Purkinje fibers. The ventricular rhythms are rather dangerous and are often treated very aggressively. The underlying causes of these rhythms are usually linked to coronary artery disease accompanied by significant ischemia or infarction, severe heart failure, cardiac valve disease, or are triggered by certain types of medications. A full review of the ventricular rhythms is beyond the scope of this book, so we will focus on two of the more common ventricular rhythms: ventricular tachycardia and ventricular fibrillation.

Ventricular Tachycardia

An example of ventricular tachycardia is shown in figure 13.10. The main criterion in ventricular rhythms is the shape of the QRS. The QRS is wide (>0.12 s in most leads) because the impulse spreads slowly through the ventricular muscle after it is generated. This increases the time it takes for the ventricles to depolarize; hence, the wide, bizarre-looking QRS. In ventricular tachycardia, this wide QRS rhythm is also very fast, with a rate of >100 beats·min^{-1}. In figure 13.10, the ventricular

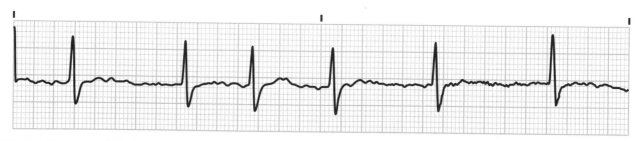

FIGURE 13.9   Atrial fibrillation.

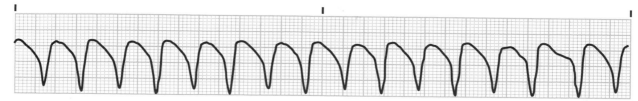

FIGURE 13.10   Ventricular tachycardia.

rate is approximately 167 beats·min⁻¹. There are no P waves that can be seen and the rhythm is usually regular. Ventricular tachycardia can start suddenly and also stop (or terminate) abruptly, even without medical intervention.

Ventricular Fibrillation

An example of ventricular fibrillation is shown in figure 13.11. The underlying physiology described in atrial fibrillation can be applied to ventricular fibrillation. Multiple areas in the ventricles are generating impulses in a very rapid and chaotic fashion. This disorganized depolarization does not elicit consistent QRS complexes leading to no effective myocardial contraction and blood ejection. Essentially, from a contractile function viewpoint, ventricular fibrillation is cardiac arrest because there is no pulse that accompanies this rhythm. Timely treatment with an automated external defibrillator (AED) or other defibrillator along with specific cardiac medications (see appendix A, Antiarrhythmic agents) is the best option.

Rhythms With "Premature Complexes"

The extra beats that are observed in some rhythms originate from specialized pacemaker cells in all parts of the heart. Everybody experiences them and they are commonly the cause of what individuals might describe as "palpitations" or "my heart skipped a beat." These beats are usually benign; however, in some situations they could be linked to underlying cardiac disease, such as coronary heart disease or myocardial infarction, especially if accompanied by other signs and symptoms of disease. In apparently healthy individuals, they can be triggered by emotional stress, intake of certain stimulants like caffeine, electrolyte imbalances, or medications that increase sympathetic nervous system activation (Goldberger, Goldberger, and Shvilkin 2013).

A common characteristic for these beats is that they are "premature," meaning that they occur earlier than the next expected beat. They can occur in a variety of ways: in random single beats that occur occasionally, in back-to-back fashion called "pairs" or "couplets," repeating patterns (i.e., occur every other beat, every third beat, or every fourth beat), and frequently (>6 per min) or infrequently (<6 per min). And, after they occur, there is a pause in the rhythm before the next normal beat occurs. Premature beats will also have very specific characteristics that are indicative of their origin; therefore, make sure to compare the premature beats to the normal beats as you determine what kind of premature beats they are. In practice, these beats are much easier to identify if you measure the R–R interval for all complexes on the rhythm strip. We will focus on two types of these extra beats: premature atrial complexes (PAC) and premature ventricular complexes (PVC).

Premature Atrial Complexes

Figure 13.12 shows an example of PACs. A PAC is an early beat that contains all the features of the cardiac cycle. A P wave is present, but it will look different than sinus rhythm P waves because this

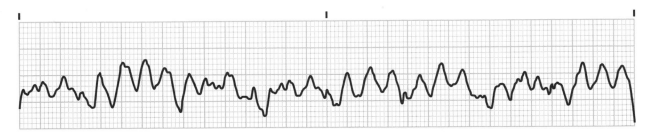

FIGURE 13.11   Ventricular fibrillation.

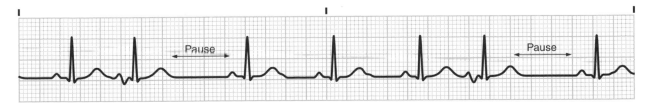

FIGURE 13.12   PACs.

beat is being conducted from a site in the atria, not the SA node. As such, the PR interval may occur outside the normal range of 0.12 to 0.20 s. The QRS will look similar to a normal beat because after the PAC is conducted, it goes through the same cardiac conduction pathway.

Premature Ventricular Complexes

Figure 13.13 shows an example of PVCs. They originate from an area within the ventricles; therefore, once the impulse is conducted, it spreads throughout the myocardium and not down the normal conduction pathways. Thus, PVCs look like the ventricular beats that were described earlier in the Ventricular Rhythms section. The QRS is wide (>0.12 s) and the T wave is pointed in the opposite direction as the QRS (i.e., QRS is positive and the T wave is negative). PVCs are usually considered more significant of a finding than PACs, especially if they accompany signs and symptoms of disease.

Basic Electrocardiography Application: Rhythm Recognition Practice

In this section of the lab, you will practice and apply your new ECG rhythm knowledge to the interpretation of basic rhythms. First, fill out the ECG rhythm recognition practice sheet with your answers for each rhythm strip in the blanks provided, and then check your interpretations against the answer key. Remember, it is important to be an "objective detective" and work through the ECG examples following the six-point interpretation checklist to arrive at your answer. Taking each problem step by step will help you correctly identify the answer and, with practice, you will be able to interpret ECGs more quickly. Have your ECG guidelines to know lab sheet next to you as you work through the rhythm strips, and enjoy the challenge!

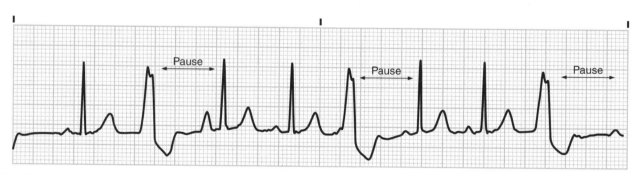

FIGURE 13.13   PVCs.

RESTING 12-LEAD ECG INTERPRETATION

Name _____

1. On your ECG, label the P wave, QRS complex, T wave, PR interval, QT interval, and ST segment for one cardiac cycle of your choosing in lead II.
2. Fill out the sheet below with the findings for your 12-lead ECG interpretation.

12-lead ECG feature	Finding	Normative value for comparison
Calibration (normal, 1/2, two times)		–
HR (bpm)		
WAVES		
P wave (amplitude and duration)		
QRS complex (duration)		
T wave (shape)		
INTERVALS		
PR interval (duration)		
QT interval (duration)		
SEGMENTS		
ST segment (position)		
Other findings of interest		
Cardiac rhythm interpretation		

Note: To complete cardiac rhythm interpretation, use either lead II on the 12-lead ECG or the rhythm strip at the bottom of your printout if your electrocardiograph provides one for you. Lead II is always good for observing cardiac rhythm because it is normally associated with the largest P waves.

From J. Janot and N. Beltz, *Laboratory Assessment and Exercise Prescription* (Champaign, IL: Human Kinetics, 2023).

ECG GUIDELINES TO KNOW LAB SHEET

ECG feature	Characteristic information
P wave	P wave is normally positive (upward) in leads I, II, aVF, aVL, V_5, V_6 P wave is negative (downward) in lead ____; biphasic in lead V_1 P wave is variable (positive or negative) in leads III, V_2–V_4 P wave is ____ s duration or ___ small squares P wave is ____ mm high or ___ small squares Biggest P waves are usually observed in lead ____
HR	Normal sinus rhythm ventricular rate is between _____ and ____ beats·min^{-1} Ventricular rate in "bradycardia" is _____ beats·min^{-1} Ventricular rate in "tachycardia" is _____ beats·min^{-1}
PR interval	Normal duration range of PR interval is ___ to ___ s or ___ to ___ small squares
QRS complex	Normal duration range of QRS complex is ___ to ___ s or ___ to ___ small squares Q wave is < ____ s duration and < ____ of its corresponding R wave Physiologic (normal) Q waves are seen in leads I, II, aVL, aVF, aVR, V_4–V_6 *Note:* Physiologic Q waves are caused by interventricular septum depolarization.
QT interval	Normal range of QT interval at rest is ___ to ___ s (dependent on HR)
ST segment	Normally isoelectric or equal with baseline in most leads
T wave	T wave is normally positive (upward) in leads I, II, aVF, V_3–V_6 T wave is negative (downward) in lead ____ T wave is variable (positive or negative) in leads III, aVL, V_1, V_2

From J. Janot and N. Beltz, *Laboratory Assessment and Exercise Prescription* (Champaign, IL: Human Kinetics, 2023).

ECG RHYTHM RECOGNITION PRACTICE SHEET

ECG Rhythm Strip 1

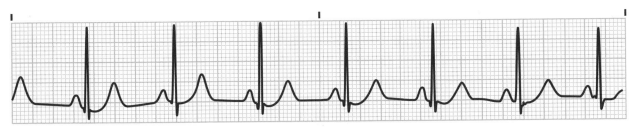

Are there P waves?	P waves present: Yes or No? Amplitude _____ mm Duration _____ s General shape characteristics _____
What is the rhythm?	Regular or irregular?
P wave and QRS complex relationship?	Each P wave followed by a QRS: Yes or No? PR interval _____ s
What is the HR?	Rate _____ beats·min^{-1}
What is the shape of the QRS?	Shape: Narrow or wide? QRS complex duration _____ s

Cardiac Rhythm Diagnosis _____

ECG Rhythm Strip 2

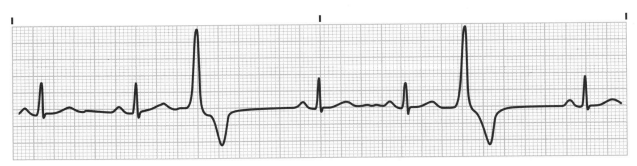

Are there P waves?	P waves present: Yes or No? Amplitude _____ mm Duration _____ s General shape characteristics _____
What is the rhythm?	Regular or irregular?
P wave and QRS complex relationship?	Each P wave followed by a QRS: Yes or No? PR interval _____ s
What is the HR?	Rate _____ beats·min^{-1}
What is the shape of the QRS?	Shape: Narrow or wide? QRS complex duration _____ s

Cardiac Rhythm Diagnosis _____

> continued

ECG Rhythm Recognition Practice Sheet   > *continued*

ECG Rhythm Strip 3

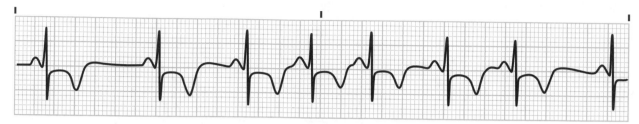

Are there P waves?	P waves present: Yes or No? Amplitude _____ mm Duration _____ s General shape characteristics _____
What is the rhythm?	Regular or irregular?
P wave and QRS complex relationship?	Each P wave followed by a QRS: Yes or No? PR interval _____ s
What is the HR?	Rate _____ beats·min⁻¹
What is the shape of the QRS?	Shape: Narrow or wide? QRS complex duration _____ s
Cardiac Rhythm Diagnosis _____	

ECG Rhythm Strip 4

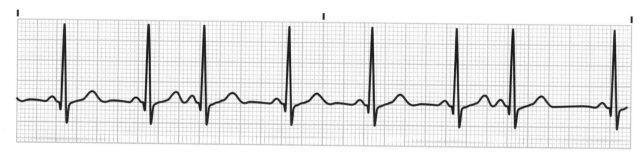

Are there P waves?	P waves present: Yes or No? Amplitude _____ mm Duration _____ s General shape characteristics _____
What is the rhythm?	Regular or irregular?
P wave and QRS complex relationship?	Each P wave followed by a QRS: Yes or No? PR interval _____ s
What is the HR?	Rate _____ beats·min⁻¹
What is the shape of the QRS?	Shape: Narrow or wide? QRS complex duration _____ s
Cardiac Rhythm Diagnosis _____	

ECG Rhythm Strip 5

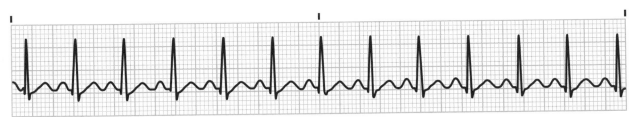

Are there P waves?	P waves present: Yes or No? Amplitude _____ mm Duration _____ s General shape characteristics _____
What is the rhythm?	Regular or irregular?
P wave and QRS complex relationship?	Each P wave followed by a QRS: Yes or No? PR interval _____ s
What is the HR?	Rate _____ beats·min⁻¹
What is the shape of the QRS?	Shape: Narrow or wide? QRS complex duration _____ s
Cardiac Rhythm Diagnosis _____	

ECG Rhythm Strip 6

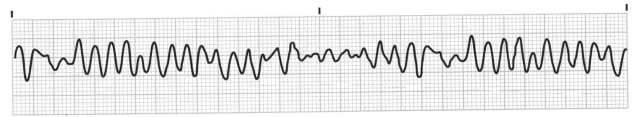

Are there P waves?	P waves present: Yes or No? Amplitude _____ mm Duration _____ s General shape characteristics _____
What is the rhythm?	Regular or irregular?
P wave and QRS complex relationship?	Each P wave followed by a QRS: Yes or No? PR interval _____ s
What is the HR?	Rate _____ beats·min⁻¹
What is the shape of the QRS?	Shape: Narrow or wide? QRS complex duration _____ s
Cardiac Rhythm Diagnosis _____	

> continued

ECG Rhythm Recognition Practice Sheet   > *continued*

ECG Rhythm Strip 7

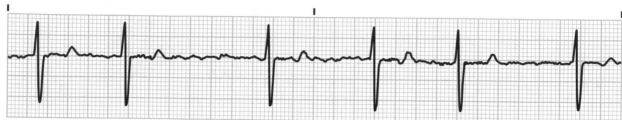

Are there P waves?	P waves present: Yes or No? Amplitude _____ mm Duration _____ s General shape characteristics _____
What is the rhythm?	Regular or irregular?
P wave and QRS complex relationship?	Each P wave followed by a QRS: Yes or No? PR interval _____ s
What is the HR?	Rate _____ beats·min⁻¹
What is the shape of the QRS?	Shape: Narrow or wide? QRS complex duration _____ s
Cardiac Rhythm Diagnosis _____	

ECG Rhythm Strip 8

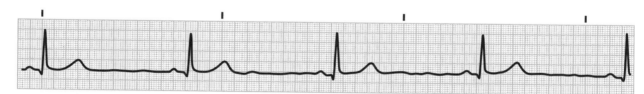

Are there P waves?	P waves present: Yes or No? Amplitude _____ mm Duration _____ s General shape characteristics _____
What is the rhythm?	Regular or irregular?
P wave and QRS complex relationship?	Each P wave followed by a QRS: Yes or No? PR interval _____ s
What is the HR?	Rate _____ beats·min⁻¹
What is the shape of the QRS?	Shape: Narrow or wide? QRS complex duration _____ s
Cardiac Rhythm Diagnosis _____	

ECG Rhythm Strip 9

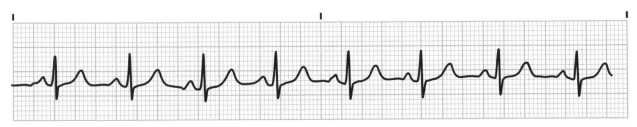

Are there P waves?	P waves present: Yes or No? Amplitude _____ mm Duration _____ s General shape characteristics _____
What is the rhythm?	Regular or irregular?
P wave and QRS complex relationship?	Each P wave followed by a QRS: Yes or No? PR interval _____ s
What is the HR?	Rate _____ beats·min⁻¹
What is the shape of the QRS?	Shape: Narrow or wide? QRS complex duration _____ s

Cardiac Rhythm Diagnosis _____

ECG Rhythm Strip 10

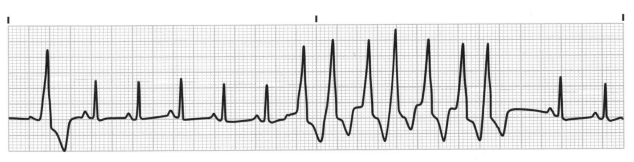

Are there P waves?	P waves present: Yes or No? Amplitude _____ mm Duration _____ s General shape characteristics _____
What is the rhythm?	Regular or irregular?
P wave and QRS complex relationship?	Each P wave followed by a QRS: Yes or No? PR interval _____ s
What is the HR?	Rate _____ beats·min⁻¹
What is the shape of the QRS?	Shape: Narrow or wide? QRS complex duration _____ s

Cardiac Rhythm Diagnosis _____

From J. Janot and N. Deltz, *Laboratory Assessment and Exercise Prescription* (Champaign, IL: Human Kinetics, 2023).

ECG RHYTHM RECOGNITION PRACTICE ANSWER KEY

ECG Rhythm Strip 1

Are there P waves?	P waves present: **Yes** Amplitude: **3.0 mm** Duration: **0.10-0.12 s** General shape characteristics: **Positive, tall, borderline wide, round on top**
What is the rhythm?	**Regular**
P wave and QRS complex relationship?	Each P wave followed by a QRS: **Yes** PR interval: **0.12-0.14 s**
What is the HR?	Rate: **72 beats·min⁻¹**
What is the shape of the QRS?	Shape: **Narrow** QRS complex duration: **0.06 s**

Cardiac Rhythm Diagnosis: **Normal sinus rhythm**

ECG Rhythm Strip 2

Are there P waves?	P waves present: **Yes** Amplitude: **2.0 mm** Duration: **0.10-0.12 s** General shape characteristics: **Positive, small, round on top, borderline wide**
What is the rhythm?	**Irregular (due to extra beats, but underlying rhythm is regular)**
P wave and QRS complex relationship?	Each P wave followed by a QRS: **Yes** PR interval: **0.18-0.20 s**
What is the HR?	Rate: **65 beats·min⁻¹** (underlying rhythm, not counting extra beats)
What is the shape of the QRS?	Shape: **Narrow** (normal beats) QRS complex duration: **0.08 s**

Cardiac Rhythm Diagnosis: **Normal sinus rhythm with PVCs[a]**

[a]In this rhythm, the underlying rhythm is normal sinus rhythm because the spacing between the normal beats is consistent. With the extra beats, which are PVCs, it makes the rhythm appear irregular. Thus, we describe the underlying rhythm first in our interpretation and then add in the premature beats. Try to establish the rate for the underlying rhythm if you can without counting the premature beats.

ECG Rhythm Strip 3

Are there P waves?	P waves present: **Yes** Amplitude: **2.0 mm** Duration: **0.12 s** General shape characteristics: **Positive, small, round on top, borderline wide**
What is the rhythm?	**Irregular**
P wave and QRS complex relationship?	Each P wave followed by a QRS: **Yes** PR interval: **0.14 s**
What is the HR?	Rate: **80 beats·min⁻¹**
What is the shape of the QRS?	Shape: **Narrow** QRS complex duration: **0.06-0.08 s**

Cardiac Rhythm Diagnosis: **Sinus arrhythmia**

ECG Rhythm Strip 4

Are there P waves?	P waves present: **Yes** Amplitude: **1.5 mm** Duration: **0.08-0.10 s** General shape characteristics: **Positive, small, narrow, and rounded in normal beats, peaked in premature beats**
What is the rhythm?	**Irregular (due to extra beats)**
P wave and QRS complex relationship?	Each P wave followed by a QRS: **Yes** PR interval: **0.12-0.14 s**
What is the HR?	Rate: **75** beats·min⁻¹ (underlying rhythm, not counting extra beats)
What is the shape of the QRS?	Shape: **Narrow** QRS complex duration: **0.10 s**

Cardiac Rhythm Diagnosis: **Normal sinus rhythm with PACs**[b]

[b]In this rhythm, the underlying rhythm is normal sinus rhythm because the spacing between the normal beats is consistent. With the extra beats, which are PACs, it makes the rhythm appear irregular. Again, describe the underlying rhythm first and then add in the premature beats. Attempt to establish the rate in the underlying rhythm if you can without these premature beats.

ECG Rhythm Strip 5

Are there P waves?	P waves present: **Yes** Amplitude: **2-2.5 mm** Duration: **0.10 s** General shape characteristics: **Borderline tall, but rounded, narrow, and positive**
What is the rhythm?	**Regular**
P wave and QRS complex relationship?	Each P wave followed by a QRS: **Yes** PR interval: **0.14-0.16 s**
What is the HR?	Rate: **130** beats·min⁻¹
What is the shape of the QRS?	Shape: **Narrow** QRS complex duration: **0.08 s**

Cardiac Rhythm Diagnosis: **Sinus tachycardia**

ECG Rhythm Strip 6

Are there P waves?	P waves present: **No** Amplitude: **NA** Duration: **NA** General shape characteristics: **NA**
What is the rhythm?	**Irregular**
P wave and QRS complex relationship?	Each P wave followed by a QRS: **NA** PR interval: **NA**
What is the HR?	Rate: **NA**
What is the shape of the QRS?	Shape: **Wide** QRS complex duration: **NA**

Cardiac Rhythm Diagnosis: **Ventricular fibrillation**[c]

[c]Most details of the cardiac cycle do not apply with this rhythm. It is more important to quickly recognize and treat this rhythm.

> continued

ECG Rhythm Recognition Practice Answer Key   > *continued*

ECG Rhythm Strip 7

Are there P waves?	P waves present: **No** Amplitude: **NA** Duration: **NA** General shape characteristics: **NA**
What is the rhythm?	**Irregular**
P wave and QRS complex relationship?	Each P wave followed by a QRS: **NA** PR interval: **NA**
What is the HR?	Rate: **60 beats·min⁻¹**
What is the shape of the QRS?	Shape: **Narrow** QRS complex duration: **0.10 s**

Cardiac Rhythm Diagnosis: **Atrial fibrillation**

ECG Rhythm Strip 8

Are there P waves?	P waves present: **Yes** Amplitude: **1.5 mm** Duration: **0.10 s** General shape characteristics: **Positive, narrow, small, and rounded**
What is the rhythm?	**Regular**
P wave and QRS complex relationship?	Each P wave followed by a QRS: **Yes** PR interval: **0.22-0.24 s**
What is the HR?	Rate: **25-30 beats·min⁻¹**
What is the shape of the QRS?	Shape: **Wide** QRS complex duration: **0.12-0.14 s**

Cardiac Rhythm Diagnosis: **Sinus bradycardia**[d]

[d]Here we have a rhythm with a longer than normal PR interval. The cause of this is an impulse conduction delay through the AV node, which is called a first-degree AV block. This condition is usually benign; thus, the HR presenting well below 60 beats·min⁻¹ remains the bigger issue in this example.

ECG Rhythm Strip 9

Are there P waves?	P waves present: **Yes** Amplitude: **2 mm** Duration: **0.10-0.12 s** General shape characteristics: **Positive, borderline wide, but small**
What is the rhythm?	**Regular**
P wave and QRS complex relationship?	Each P wave followed by a QRS: **Yes** PR interval: **0.16 s**
What is the HR?	Rate: **83 beats·min⁻¹**
What is the shape of the QRS?	Shape: **Narrow** QRS complex duration: **0.06-0.08 s**

Cardiac Rhythm Diagnosis: **Normal sinus rhythm**

ECG Rhythm Strip 10

Are there P waves?	P waves present: **Yes** Amplitude: **2-2.5 mm** Duration: **0.08-0.10 s** General shape characteristics: **Positive, borderline tall, but narrow**
What is the rhythm?	**Irregular (due to extra beats)**
P wave and QRS complex relationship?	Each P wave followed by a QRS: **Yes (with normal beats)** PR interval: **0.14-0.16 s**
What is the HR?	Rate: **100** beats·min⁻¹ for underlying rhythm; **150** beats·min⁻¹ for second rhythm
What is the shape of the QRS?	Shape: **Narrow** for underlying rhythm; **wide** for second rhythm QRS complex duration: **0.06** s for underlying rhythm; **0.16 s** for second rhythm

Cardiac Rhythm Diagnosis: **Normal sinus rhythm with ventricular tachycardia and a single PVC**[e]

[e]In this rhythm, the underlying rhythm is normal sinus rhythm (borderline rate) with a second rhythm of ventricular tachycardia. The ventricular tachycardia lasts for 7 beats and then stops. This can be described as a "7-beat run of ventricular tachycardia."

From J. Janot and N. Beltz, *Laboratory Assessment and Exercise Prescription* (Champaign, IL: Human Kinetics, 2023).

Assessment of Clinical Variables: Functional Fitness Testing

Purpose of the Lab

With the growing population of older adults in the United States, there has been an increased emphasis on training young professionals to work with individuals aged 65 and older. In 2016, the U.S. population of older adults represented 15.2% (49.2 million) of the total population, and this number is expected to grow to 23.4% (94.7 million) by 2060 (United States Census Bureau 2018). To understand the underlying impact of the aging population, we first need to examine the concept of frailty and its effect on disability through the aging process. Frailty can be defined as a state of muscular weakness and other secondary widely distributed losses in function and structure that are usually initiated by decreased levels of physical activity (Bortz 2002). Although frailty is a multisystem concept, the constant is the interplay between physical activity, muscle weakness, and a diminished ability to perform activities of daily living (ADLs) and instrumental activities of daily living (IADLs) both safely and independently. Examples of ADLs include selecting and putting on clothes, the ability to groom and bathe, and ambulation from one position to another, while IADLs include things like grocery shopping, paying bills, cleaning the house, and preparing meals (Edemekong et al. 2021). Frailty can be thought of as a process occurring over time due to diminishing physiological reserve capacity (Campbell and Buchner 1997). It is easiest to think of physiological reserve capacity as a "buffer zone" in ability to complete ADLs and IADLs at peak physiological performance and the inability to complete them safely, independently, and without undue fatigue. For example, most individuals reach the peak of their physical development between 20 and 30 years old. Key components over the lifespan such as genetics, disease and injuries, lifestyle, and the aging process cause a loss in musculoskeletal function, aerobic fitness, cognitive ability, and nutritional status. The diminished physical components reduce an individual's physiological reserve capacity, making ADLs and IADLs more difficult and reducing functional abilities. By the time the sixth and seventh decades of life have been reached, an individual may only possess 20% to 30% of the peak physical performance that they possessed in their twenties.

The impact of losing physiological reserve capacity can be better understood by examining a specific task. For example, walking 3.0 mi·h⁻¹ during your physiological prime only represents a very small portion of your reserve capacity. The net result is a low perceived effort during that task. As you lose reserve capacity, the task of walking 3.0 mi·h⁻¹ represents an intensity closer to your maximum ability and is perceived as more difficult. A threshold of 30% of maximal function has been suggested as the cutoff for independent and dependent living, giving us a 70% buffer in physiological reserve capacity before reaching disability.

We will be assessing functional fitness performance in this lab. Rikli and Jones (1999) define functional fitness as having the physiologic capacity to perform normal everyday activities safely and independently without undue fatigue. As we discussed earlier, the process of frailty is a dynamic process that occurs with or without the presence of disease and may eventually lead to disability. A major goal of a practitioner working with older adults is to identify physical impairments that may contribute to functional limitations and reductions in ability to complete ADLs or IADLs (Rikli and Jones 1997).

For any ADL or IADL, there is a function or set of functions that require some base level of physical performance. For example, doing the laundry may require an individual to climb stairs, carry a laundry basket, and bend over to add and remove clothing from a washer and dryer. Stair climbing requires lower body strength and endurance, carrying laundry requires upper body strength, and bending requires lower body and lower back flexibility. Rikli and Jones (1999) tested 7,183 individuals across 267 sites and 21 states to measure physical performance in older adults and develop the functional fitness testing (FFT) battery. The specific physical parameters assessed within the FFT are upper body strength, lower body strength,

aerobic endurance, agility and dynamic balance, upper body flexibility, and lower body flexibility. The individual FFT tasks and their associated physical parameter are listed as follows:

- 30-second chair stand: Lower body strength
- 30-second arm curl: Upper body strength
- 6-minute walk test: Aerobic endurance
- 2-minute step test: Aerobic endurance (alternative test)
- 8-foot up-and-go: Agility and dynamic balance
- Chair sit-and-reach: Lower body flexibility
- Back scratch: Upper body flexibility

It is important to understand that the specific tests used in the FFT are not reference criteria methods but rather adapted for suitability in an older population. For example, we learned that the one-repetition maximum (1RM) bench press test is the reference criterion to evaluate upper body strength; however, it may not be appropriate to complete a 1RM test in older individuals or those who are frail. Even so, the assessments used in the FFT have shown to be valid estimates (r = 0.73-0.83) compared to their reference criterion measurements (Rikli and Jones 1999).

Measuring functional fitness can help identify clients who are becoming increasingly frail and are at risk for disability. The entire FFT battery takes only ~20 to 30 min, allowing for quick reevaluation to track progress through an exercise program. The comprehensive nature of testing helps practitioners plan exercise programs specific to targeting areas of weakness. For example, a client may have impairments specific to shoulder and lower body flexibility but perform well on all other FFT assessments. The evaluation of FFT performance is an opportunity to apply theoretical foundations learned throughout previous labs into a single lab experience. Therefore, the practitioner can emphasize the importance of maintaining muscular strength, aerobic endurance, flexibility, agility, and dynamic balance and use it to set goals and motivate clients.

Another major consequence of frailty is the increased risk of falls. Falls are the leading injury-related cause of death in older individuals, affecting 28.7% of community-dwelling older adults in 2014 (Bergen, Stevens, and Burns 2016).

Additionally, falls account for 87% of fractures in older adults (Ambrose, Cruz, and Paul 2015). Fractures have been shown to be a predictor of long-term mortality in older patients. For example, an older individual has a 27% increased risk of dying within the year following a hip fracture (Cenzer et al. 2016). Although fall prediction involves a complex system of factors, lower body muscular strength, balance, and self-selected gait speed have been established as independent predictors of falls (Kyrdalen et al. 2019; Nowak and Hubbard 2009). The balance error scoring system (BESS) is a quick, inexpensive, and easy test to assess static balance and postural stability. The BESS is comprised of three different stances (narrow double leg, single leg, and tandem) tested on two surfaces (hard floor and medium-density foam) for a total of six total testing conditions. A point is given for errors during each testing condition and then totaled as a composite score (Riemann, Guskiewicz, and Shields 1999).

Combined with FFT and BESS, the measurement of gait speed (GS) is an additional rapid and easy assessment to indicate functional ability. Despite its simplicity, research has shown the utility of GS to predict functional dependence, disability, hospitalization, and even all-cause mortality (Middleton, Fritz, and Lusardi 2015). Altogether, combining FFT, BESS, and GS testing will give you a robust lab experience to indicate functional ability and fall risk in older adults. The purpose of this lab is to complete FFT, BESS, and GS. You will use these data to evaluate your client's functional ability, balance, and gait speed. Finally, we will discuss current balance exercise prescription training principles and you will design a progressive balance exercise prescription.

Necessary Equipment or Materials

- Chair (stable, non-rolling)
- Stopwatch
- 8 lb dumbbell
- 5 lb dumbbell
- Cones
- 20 yd × 5 yd area (a 50 yd hallway is another option)

- Tape measure (Gulick preferred)
- 25 ft tape measure
- Medium-density foam pad
- Functional fitness testing data sheet
- Balance exercise progression planning sheet

Calibration of Equipment

For this lab, no calibration of equipment is needed. Make sure that all equipment is in good working condition before use.

Procedures
Functional Fitness Testing Battery
30-Second Chair Stand

1. Place the back of the chair against a wall to ensure that it will not move during the testing procedure.
2. Instruct the client to start from a seated position with feet shoulder-width apart and arms over their chest (figure 14.1a).
3. Instruct the client to rise to a standing position (figure 14.1b), then lower back into a seated position. The client must reach full extension of the knees and hips in the standing position, and their buttocks must touch the seat of the chair for a valid repetition. Encourage the client to complete as many repetitions as possible from seated to standing position in 30 s.
4. Demonstrate proper technique and allow the client to ask questions.
5. Start the stopwatch on the client's first movement.
6. Count each repetition as the client reaches the bottom position of the chair stand.
7. Upon reaching 30 s, instruct the client to stop.
8. Record the value on the lab 14 data collection sheet.
9. Interpret results using table 14.1.

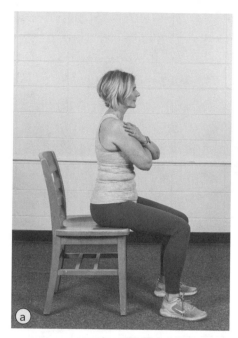

FIGURE 14.1   30-second chair stand: *(a)* start; *(b)* finish.

TABLE 14.1   30-Second Chair Stand Norms (Repetitions) by Sex and Age Group

Age (yr)	PERCENTILE								
	10	20	30	40	50	60	70	80	90
MEN									
60-64	11	13	14	15	16	17	19	20	22
65-69	9	11	13	14	15	16	18	19	21
70-74	9	11	12	13	14	16	17	18	20
75-79	8	10	12	13	14	15	16	18	20
80-84	7	9	10	11	12	13	14	16	17
85-89	5	7	9	10	11	12	13	15	17
90-94	5	7	8	9	10	11	12	13	15
WOMEN									
60-64	9	11	12	14	15	16	17	18	20
65-69	9	11	12	13	14	14	15	16	18
70-74	8	10	11	12	13	14	15	16	18
75-79	8	9	11	12	12	13	14	16	17
80-84	6	8	9	10	11	12	13	15	17
85-89	5	7	8	9	10	11	12	14	15
90-94	1	4	5	7	8	9	11	12	15

Adapted by permission from R.E. Rikli and C.J. Jones, *Senior Fitness Test Manual*, 2nd ed. (Champaign, IL: Human Kinetics, 2013), 154.

30-Second Arm Curl

1. Instruct the client to start from a seated position, sitting upright with their back against the chair.

2. Ask the client which arm they prefer to test. The arm curl test only requires testing of a single arm, because discrepancies in strength shouldn't be significant between arms.

3. If the client is a woman, use the 5 lb dumbbell. If the client is a man, use the 8 lb dumbbell.

4. The client should start with the dumbbell in the extended position (figure 14.2a). Instruct the client to flex the elbow to raise the dumbbell toward the shoulder (figure 14.2b), then lower to the extended position. They must reach complete flexion of the elbow and then return to the extended position for a valid repetition. Encourage the client to complete as many repetitions as possible in 30 s.

5. Demonstrate proper technique and allow the client to ask questions.

6. Start the stopwatch on the client's first movement.

7. Count each repetition as the client reaches the extended position.

8. Upon reaching 30 s, instruct the client to stop.

9. Record the value on the lab 14 data collection sheet.

10. Interpret results using table 14.2.

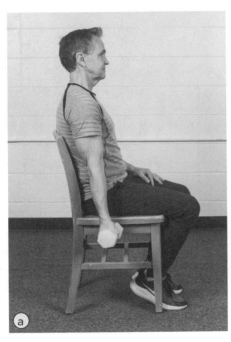
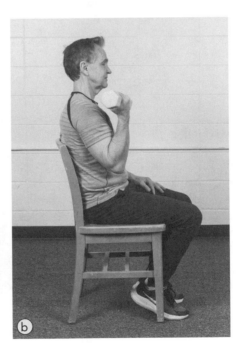

FIGURE 14.2   30-second arm curl: (a) start; (b) finish.

TABLE 14.2   30-Second Arm Curl Norms (Repetitions) by Sex and Age Group

Age (yr)	PERCENTILE								
	10	20	30	40	50	60	70	80	90
MEN									
60-64	13	15	17	18	19	20	21	23	25
65-69	12	14	16	17	18	20	21	23	25
70-74	11	13	15	16	17	19	20	22	24
75-79	10	12	14	15	16	17	19	20	22
80-84	10	12	14	15	16	17	18	20	22
85-89	8	10	11	13	14	15	16	17	19
90-94	8	9	10	11	12	13	14	15	16
WOMEN									
60-64	10	12	14	15	16	17	18	20	22
65-69	10	12	13	14	15	16	17	19	21
70-74	9	11	12	13	15	16	17	18	20
75-79	8	10	12	13	14	15	16	18	20
80-84	8	10	11	12	13	14	15	16	18
85-89	7	9	10	11	12	13	14	15	17
90-94	6	8	9	10	11	12	13	14	16

Adapted by permission from R.E. Rikli and C.J. Jones, *Senior Fitness Test Manual*, 2nd ed. (Champaign, IL: Human Kinetics, 2013), 155.

Chair Sit-and-Reach

1. Instruct the client to sit toward the front half of the chair and fully extend one leg with their foot in full dorsiflexion. They should have their other leg bent at 90 degrees and foot flat on the ground.

2. Client should then reach down as far as possible toward the dorsiflexed toe with overlapping hands (figure 14.3).

3. Demonstrate proper technique and allow the client to ask questions.

4. Use the tape measure to measure the distance from the tip of the closest finger to the toe.

5. An inability to reach the toe results in a negative score (in inches from the toe), reaching the toe results in a score of zero, and reaching past the toe results in a positive score (in inches past the toe).

FIGURE 14.3   Chair sit-and-reach.

6. Alternate the measurement procedure between legs until two trials have been completed with each leg.

7. Record scores for left and right legs on the lab 14 data collection sheet. Use the best score for each in your overall interpretation.

8. Interpret results using table 14.3.

TABLE 14.3   Chair Sit-and-Reach Norms (Inches) by Sex and Age Group

Age (yr)	PERCENTILE								
	10	20	30	40	50	60	70	80	90
MEN									
60-64	−5.5	−3.4	−1.9	−0.6	0.6	1.8	3.1	4.6	6.7
65-69	−5.9	−3.9	−2.4	−1.1	0.0	1.1	2.4	3.9	5.9
70-74	−5.9	−3.9	−2.4	−1.2	0.0	1.1	2.4	3.8	5.8
75-79	−7.1	−5.0	−3.5	−2.3	−1.1	−0.1	1.3	2.8	4.9
80-84	−8.4	−6.2	−4.6	−3.2	−2.0	−0.8	0.6	2.2	4.4
85-89	−7.8	−5.9	−4.6	−3.5	−2.4	−1.3	−0.2	1.1	3.0
90-94	−9.1	−7.2	−5.8	−4.7	−3.6	−2.5	−1.4	0.0	1.9
WOMEN									
60-64	−3.0	−1.3	0.0	1.1	2.1	3.1	4.2	5.5	7.2
65-69	−2.6	−1.0	0.1	1.1	2.0	2.9	3.9	5.0	6.6
70-74	−3.3	−1.7	−0.5	0.5	1.4	2.3	3.3	4.5	6.1
75-79	−3.7	−2.0	−0.8	0.2	1.2	2.1	3.2	4.4	6.1
80-84	−4.2	−2.6	−1.4	−0.4	0.5	1.4	2.4	3.6	5.2
85-89	−4.8	−3.2	−2.0	−1.0	−0.1	0.8	1.8	3.0	4.6
90-94	−6.8	−5.1	−3.8	−2.7	−1.7	−0.7	0.4	1.7	3.4

Adapted by permission from R.E. Rikli and C.J. Jones, *Senior Fitness Test Manual*, 2nd ed. (Champaign, IL: Human Kinetics, 2013), 158.

Back Scratch

1. The client will complete all trials from a standing position, if able.

2. Instruct the client to reach back and behind the head with the top hand, moving the hand down between the shoulder blades with the palm against their back. The opposite, bottom hand should reach behind and up the back with palm facing outward (figure 14.4).

3. Demonstrate proper technique and allow the client to ask questions.

4. Use the tape measure to measure the distance between the fingertips of the two middle fingers.

5. An inability to touch the fingertips results in a negative score (in inches from opposing fingertips), touching fingertips results in a score of zero, and overlapping the fingertips results in a positive score (in inches overlapping).

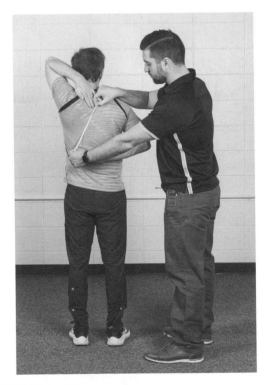

FIGURE 14.4   Back scratch.

6. Alternate the measurement procedure by switching top and bottom hands until two trials have been completed with each combination.

7. Record scores as left-over-right (LoR) and right-over-left (RoL) on the lab 14 data collection sheet. Use the best score for each in your overall interpretation.

8. Interpret results using table 14.4.

TABLE 14.4   Back Scratch Norms (Inches) by Sex and Age Group

Age (yr)	PERCENTILE								
	10	20	30	40	50	60	70	80	90
MEN									
60-64	−9.5	−7.4	−5.9	−4.6	−3.4	−2.2	−0.9	0.6	4.5
65-69	−10.4	−8.2	−6.6	−5.3	−4.1	−2.9	−1.6	0.0	2.2
70-74	−10.8	−8.6	−7.0	−5.7	−4.5	−3.3	−2.0	−0.4	1.8
75-79	−12.1	−9.9	−8.3	−6.9	−5.6	−4.3	−2.9	−1.3	0.9
80-84	−12.6	−10.2	−8.5	−7.1	−5.7	−4.3	−2.9	−1.2	1.2
85-89	−12.3	−10.2	−8.7	−7.4	−6.2	−5.0	−3.7	−2.2	−0.1
90-94	−13.3	−11.2	−9.7	8.4	−7.2	−6.0	−4.7	−3.2	−1.1
WOMEN									
60-64	−5.2	−3.6	−2.5	−1.6	−0.7	0.2	1.1	2.2	3.8
65-69	−5.9	−4.3	−3.1	−2.1	−1.2	−0.3	0.7	1.9	3.5
70-74	−6.6	−4.9	−3.7	−2.6	−1.7	−0.8	0.3	1.5	3.2
75-79	−7.3	−5.5	−4.2	−3.1	−2.1	−1.1	0.0	1.3	3.1
80-84	−8.0	−6.1	−4.8	−3.7	−2.6	−1.6	−0.4	0.9	2.8
85-89	−9.7	−7.7	−6.2	−5.0	−3.9	−2.8	−1.6	−0.1	1.9
90-94	−11.2	−8.9	−7.2	−5.8	−4.5	−3.2	−1.8	−0.1	2.2

Adapted by permission from R.E. Rikli and C.J. Jones, *Senior Fitness Test Manual*, 2nd ed. (Champaign, IL: Human Kinetics, 2013), 159.

8-Foot Up-and-Go

1. Place the back of the chair against a wall to ensure that it will not move during the testing procedure.

2. Using a 25 ft tape measure, measure out 8 ft from the front of the chair. Place a cone on the ground at the 8 ft mark.

3. Instruct the client to start from a seated position.

4. The client's objective is to stand, walk 8 ft, walk around the cone, and return to a seated position. Encourage the client to complete the trial as fast as possible.

5. Demonstrate proper pathing and allow the client to ask questions.

6. Start the stopwatch when the client begins to move from the seated position.

7. Stop the stopwatch once the client has returned to a seated position on the chair.

8. Complete the trial three times. Record all times on the lab 14 data collection sheet. Use the fastest time in your overall interpretation.

9. Interpret results using table 14.5.

TABLE 14.5 8-Foot Up-and-Go Norms (Seconds) by Sex and Age Group

	PERCENTILE								
Age (yr)	10	20	30	40	50	60	70	80	90
MEN									
60-64	6.4	5.8	5.4	5.0	4.7	4.4	4.0	3.6	3.0
65-69	6.6	6.1	5.7	5.4	5.1	4.8	4.5	4.1	3.6
70-74	7.0	6.4	6.0	5.6	5.3	5.0	4.6	4.2	3.6
75-79	8.3	7.5	6.9	6.4	5.9	5.4	4.9	4.3	3.5
80-84	8.7	7.9	7.3	6.9	6.4	6.0	5.5	4.9	4.1
85-89	10.5	9.4	8.6	7.9	7.2	6.5	5.8	5.0	4.3
90-94	11.8	10.5	9.6	8.8	8.1	7.4	6.6	5.7	4.5
WOMEN									
60-64	6.7	6.2	5.8	5.5	5.2	4.9	4.6	4.2	3.7
65-69	7.1	6.6	6.2	5.9	5.6	5.3	5.0	4.6	4.1
70-74	8.0	7.3	6.8	6.4	6.0	5.6	5.2	4.7	4.0
75-79	8.3	7.6	7.1	6.7	6.3	5.9	5.5	5.0	4.3
80-84	10.0	9.0	8.3	7.8	7.2	6.7	6.1	5.4	4.4
85-89	11.1	10.0	9.2	8.5	7.9	7.3	6.6	5.8	4.7
90-94	13.5	12.1	11.1	10.2	9.4	8.6	7.7	6.7	5.3

Adapted by permission from R.E. Rikli and C.J. Jones, *Senior Fitness Test Manual*, 2nd ed. (Champaign, IL: Human Kinetics, 2013), 160.

6-Minute Walk Test

1. If space allows, measure out a 20 yd × 5 yd rectangular course (figure 14.5). Using a linear 50 yd course will also work. 1 yd = 0.9144 m = 3 ft.

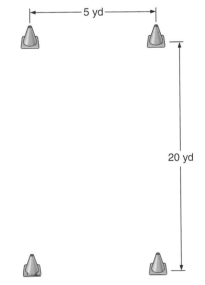

FIGURE 14.5   6-minute walk test course.

2. The client's objective is to walk around the outside of the 50 yd rectangular course, completing as many laps as possible, for a duration of 6 min. If using a linear course, the client will walk down 50 yd, turn around, and return to the starting position as many times as possible for the duration of the test. Encourage the client to walk as fast as possible during the test.

3. Demonstrate proper pathing and allow the client to ask questions.

4. Start the stopwatch on the client's first movement.

5. Record the number of laps completed around the 50 yd rectangular course or 50 yd segments completed on the linear course.

6. Upon reaching 6 min, instruct the client to stop.

7. Add any partial laps or segments completed to your tally. Add up the total distance covered (in yards) and record on your lab 14 data collection sheet.

8. Interpret results using table 14.6.

TABLE 14.6   6-Minute Walk Test Norms (Yards) by Sex and Age Group

Age (yr)	PERCENTILE								
	10	20	30	40	50	60	70	80	90
MEN									
60-64	556	597	626	651	674	697	722	751	792
65-69	499	544	577	605	631	657	685	718	763
70-74	481	526	559	586	612	638	665	698	743
75-79	394	449	489	542	555	586	621	661	716
80-84	370	423	462	494	524	554	586	625	678
85-89	295	358	403	442	477	512	551	596	659
90-94	214	279	326	366	403	440	480	527	592
WOMEN									
60-64	495	532	559	582	603	624	647	674	711
65-69	439	483	515	543	568	593	621	653	697
70-74	423	466	497	524	548	572	599	630	673
75-79	363	413	450	480	509	538	568	605	655
80-84	312	364	401	433	462	491	523	560	612
85-89	261	318	359	394	426	458	493	534	591
90-94	196	251	291	326	357	388	423	463	518

Adapted by permission from R.E. Rikli and C.J. Jones, *Senior Fitness Test Manual*, 2nd ed. (Champaign, IL: Human Kinetics, 2013), 156.

2-Minute Step Test

This test should be used when the 6-minute walk test cannot be completed due to space or weather limitations.

1. The client must complete the trial from a standing position.

2. The client's objective is to march in place for the 2 min test duration. The knee should reach a height equal to the distance of half-way between the base of the patella and iliac crest during each step.

3. To standardize the step height, instruct the client to place their hands on the front of their thighs with straight arms. The client should then flex 30 degrees at the shoulder, moving the hands slightly in front of the body. This should represent an appropriate height for the test.

4. During stepping, the knee should touch the ipsilateral hand each step (figure 14.6). Ensure that the client's hands do not move upward or downward during the 2 min duration. Encourage the client to step as fast as possible during the test.

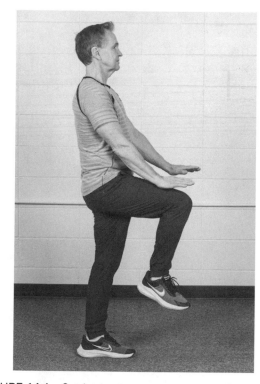

FIGURE 14.6   2-minute step test.

5. Demonstrate proper technique and allow the client to ask questions.

6. Start the stopwatch on the client's first movement.

7. Count and record the number of times the right knee touches the right hand during the trial on the lab 14 data collection sheet.

8. Interpret results using table 14.7.

TABLE 14.7   2-Minute Step Test Norms (Steps) by Sex and Age Group

Age (yr)	PERCENTILE								
	10	20	30	40	50	60	70	80	90
MEN									
60-64	74	83	90	96	101	106	112	119	128
65-69	72	82	89	95	101	107	113	120	130
70-74	66	76	83	89	95	101	107	114	124
75-79	56	68	77	84	91	98	105	114	126
80-84	56	67	75	81	87	93	99	107	118
85-89	44	55	63	69	75	81	87	95	106
90-94	36	47	55	62	69	76	83	91	102
WOMEN									
60-64	60	71	79	85	91	97	103	111	122
65-69	57	68	76	84	90	96	104	112	123
70-74	52	63	71	78	84	90	97	105	116
75-79	53	64	72	78	84	90	96	104	115
80-84	46	56	63	69	75	81	87	94	104
85-89	42	52	59	64	70	76	81	88	98
90-94	31	40	47	53	58	63	69	76	85

Adapted by permission from R.E. Rikli and C.J. Jones, *Senior Fitness Test Manual*, 2nd ed. (Champaign, IL: Human Kinetics, 2013), 157.

Balance Error Scoring System

The BESS consists of three stances tested on two different surfaces for a total of six trials. Each stance is performed with hands on the iliac crests and eyes closed for 20 s. All trials are performed barefoot or wearing socks. Demonstrate each trial and allow the client to ask questions.

1. Narrow double-leg stance on floor: Place feet together, side by side (figure 14.7a).

2. Single-leg stance on floor: Stand on nondominant foot with hip in 30-degree flexion and knee in 45-degree flexion (figure 14.7b).

3. Tandem stance on floor: Place nondominant foot behind dominant foot, ensuring that the toes of the nondominant foot are touching the heel of the dominant foot (figure 14.7c).

4. Narrow double-leg stance on medium-density foam pad (figure 14.8a).

5. Single-leg stance on medium-density foam pad (figure 14.8b).

6. Tandem stance on medium-density foam pad (figure 14.8c).

During each trial, keep score by adding a point for each of the following errors committed:

- Opening eyes
- Stepping, stumbling, or falling
- Lifting heel or forefoot
- Lifting hands from the hips
- Further flexion of the hip during single-leg stance trial
- If the client remains out of test position for more than 5 s, assign an error score of 10 and end the trial.

Record the error score for each trial on the lab 14 data collection sheet. Add the error scores for all six trials and record as a total composite score on the lab 14 data collection sheet. Interpret your results using table 14.8.

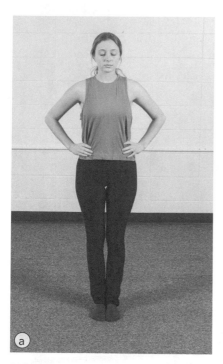

FIGURE 14.7   BESS floor trials: *(a)* narrow double leg; *(b)* single leg; *(c)* tandem.

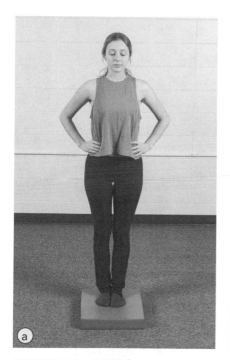

FIGURE 14.8   BESS foam mat trials: *(a)* narrow double leg; *(b)* single leg; *(c)* tandem.

TABLE 14.8   Balance Error Scoring System Norms by Sex and Age Group

Age (yr)	CATEGORY					
	Very poor	Poor	Below average	Normal	Above average	Superior
MEN						
20-29	≥22	16-21	15	7-14	5-6	0-4
30-39	≥27	19-26	16-18	7-15	5-6	0-4
40-49	≥28	21-27	17-20	8-16	6-7	0-5
50-54	≥29	24-28	18-23	8-17	7	0-6
55-59	≥35	29-34	21-28	11-20	8-10	0-7
60-64	≥36	28-35	22-27	12-21	9-11	0-8
65-69	≥40	34-39	24-33	15-23	13-14	0-12
WOMEN						
20-29	≥26	20-25	15-19	8-14	6-7	0-5
30-39	≥28	20-27	16-19	7-15	5-6	0-4
40-49	≥30	21-29	16-20	8-15	6-7	0-5
50-54	≥36	25-35	21-24	10-20	8-9	0-7
55-59	≥40	29-39	22-28	11-21	9-10	0-8
60-64	≥44	32-43	23-31	13-22	10-12	0-9
65-69	≥39	28-38	25-27	15-24	14	0-13

Adapted using data from G.L. Iverson and M.S. Koehle, "Normative Data for the Balance Error Scoring System in Adults," *Rehabilitation Research and Practice* 2013: 846418, which is distributed under Creative Commons Attribution License CC-BY (https://creativecommons.org/licenses/by/4.0/).

4-Meter Gait Speed

1. Place cones to mark the starting line for the test: one cone at 2 m from the starting line, and the second cone at 6 m from the starting line (figure 14.9).

2. Instruct the client to start the test with their shoes behind the starting line. Tell them to walk at their usual walking speed, as if they were walking down the street.

3. Demonstrate the procedure and allow the client to ask questions.

4. Give a countdown command of "3, 2, 1, go!"

5. Start the stopwatch as soon as the client's front foot crosses the 2 m line and stop it once the front foot crosses the 6 m finish line.

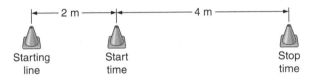

FIGURE 14.9   4-meter gait speed course.

6. Allow the client to complete three total trials. Record each trial on the lab 14 data collection sheet. Calculate average trial time.

7. Convert to m·s⁻¹ by dividing 4 m by their average trial time. Record as average speed on the lab 14 data collection sheet.

8. Interpret the results using table 14.9.

TABLE 14.9   4-Meter Gait Speed Means (m·s⁻¹) ± Standard Deviations by Sex and Age Group

Age (yr)	Speed
MEN	
18-29	1.18 ± 0.20
30-39	1.21 ± 0.21
40-49	1.21 ± 0.24
50-59	1.16 ± 0.20
60-69	1.16 ± 0.22
70-79	1.07 ± 0.24
80-85	0.97 ± 0.20
WOMEN	
18-29	1.11 ± 0.20
30-39	1.15 ± 0.20
40-49	1.14 ± 0.23
50-59	1.15 ± 0.22
60-69	1.05 ± 0.22
70-79	0.99 ± 0.22
80-85	0.95 ± 0.24

Data from Bohannon and Wang (2019).

Normative Data and Interpretation

Functional Fitness Testing Battery

Interpret your client's individual FFT scores by comparing their scores to those in their respective norm chart (tables 14.1-14.7). Remember to make the interpretation relevant and meaningful. For example, your 66-year-old female client has an average 8-foot up-and-go time of 6.1 s. Their agility and dynamic balance are between the thirtieth and fortieth percentile. This means that they are in the bottom 30% to 40% for a female their age, or that 60% to 70% of females their age perform better on the 8-foot up-and-go.

Balance Error Scoring System

Interpret the BESS as the total composite score of all six trials using table 14.8.

Gait Speed

Interpret your client's gait speed as either "above average" or "below average" according to table 14.9. The most significant clinical interpretation of gait speed is that a walking speed of 1.0 m·s⁻¹ can be used as a threshold value to predict an individual's ability to maintain independence and makes them less likely to be hospitalized or have an adverse health event. It also can predict their overall success as a safe ambulator in the community (Fritz and Lusardi 2009).

Exercise Program Design: Balance Training Exercise Prescription

Currently, optimal prescription recommendations for balance training in older adults are largely unknown. We know that regular physical activity consisting of at least two components of strength, aerobic, flexibility, and balance training can reduce the risk of fall-related injuries by 32% to 40% (ACSM 2021). The concept of neuromotor training has arisen to describe a combination of balance, coordination, gait, agility, and proprioceptive skills derived from various resistance, flexibility, and balance exercises. Therefore, the current recommendation for the most effective means to

Case Study for Progressive Balance Exercise Prescription

For the program design, you will fill in the balance exercise progression planning sheet. For type of balance exercise, you will progress through basic, moderate, and advanced variations of sitting, standing, and walking exercises. Remember, timing of the program can be prior to, during, or after a resistance training or aerobic training session. For simplicity, this prescription is for a client who is new to balance exercises. For examples of select balance exercises, review chapter 12 in the *Advanced Fitness Assessment and Exercise Prescription* text by Gibson, Wagner, and Heyward (2019). Balance programming doesn't have rigid guidelines like resistance, aerobic, and flexibility training, but this allows us to add creativity to our programming. Always ensure that your client can perform the exercises safely, and provide assistance if needed.

Demography

Age: 68
Height: 5 ft 1 in. (154.9 cm)
Weight: 126 lb (57.2 kg)
Sex: Female
Race or Ethnicity: Southeast Asian

Family History

Your client's family history reveals a history of cardiovascular disease. Their mother passed away at 89 yrs of age from a stroke, and their father passed away at 81 yrs of age from heart failure.

Medical History

Present Conditions

On the health history questionnaire form, your client reports a history of multiple sclerosis, diagnosed at 51 years old. The most common side effects reported are balance and coordination issues and weakness, particularly during the warmer seasons. Typically, your client feels better and experiences fewer symptoms during the colder winter months. Your client takes Tylenol for mild arthritic joint pain, as needed. A recent physical assessment revealed the following health information. Your client's resting HR was 74 beats·min^{-1} and resting BP was 114/66 mmHg (confirmed from a previous measurement). The client's BIA assessment showed a percent body fat (%BF) of 36%. Fasting blood glucose was 97 mg·dL^{-1}. No blood chemistry panel for cholesterol is reported at this time. Their current exercise test, which was the Ebbeling walk test, showed a $\dot{V}O_2$max value of 22.7 mL·kg^{-1}·min^{-1}. Their FFT, BESS, and GS test results follow.

Functional fitness testing		Balance error scoring system
30-second chair stand	14 reps	Composite score: 25
30-second arm curl	18 reps	
Chair sit-and-reach	L: −0.5 in. R: 1.5 in.	
Back scratch	LoR: −2.5 in. RoL: 0.0 in.	
8-foot up-and-go	5.27 s	**Gait speed**
6-minute walk test	590 yd	Speed: 0.97, 1.02, 1.05 m·s^{-1}

Past Conditions

Otherwise, your client reports no other past medical and health issues.

> *continued*

Case Study for Progressive Balance Exercise Prescription   > *continued*

Behavior and Risk Assessment

Your client participates in regular physical activity by walking briskly 45 to 60 min every day. They comment on how regular exercise reduces their multiple sclerosis symptoms. They have been retired for 5 yr and previously worked as a pediatric nurse for 40 yr. Your client is coming to you because they have been struggling with their balance and coordination. You explain that the best way to improve balance is to engage in a well-rounded exercise program consisting of resistance, aerobic, and balance training. You feel comfortable with your client's current aerobic exercise plan, adding some occasional treadmill walking to increase intensity. The program that you have designed includes 3 d·wk^{-1} of moderate intensity resistance training for 10 wk. Their main goals for the balance program are to improve their mobility and balance so they have improved self-efficacy and worry less about falls as they age. Your client wants to integrate the balance program within the resistance training program to utilize time and equipment at the gym.

Review your client's data and provide a rating for each of the assessments in the functional fitness testing worksheet data table. Also, please fill out the planning sheet with progressions for sitting, standing, and walking balance exercises.

maintain or improve balance and coordination is through a combination of aerobic, resistance, and flexibility prescription. Bull and colleagues (2020) strongly recommend physical activity that emphasizes functional balance and strength training on three or more days per week in older adults. Specific balance training involves static or dynamic exercises to challenge postural sway on a stable or unstable surface. These exercises should be added to a regular aerobic, resistance, and flexibility training program. Balance exercises can be integrated within the warm-up, as a main component of the exercise prescription, or in the cool-down. Other general guidelines for balance training are that it includes the following components (ACSM 2021):

- Progression in postures that reduce the base of support
- Dynamic movements that perturb the center of gravity
- Stressing postural muscle groups
- Reducing sensory input

Three easy categories of balance exercise prescription are sitting, standing, and dynamic exercises. Once you have established a starting exercise, you can progress the exercise to make it more difficult. Examples of each are as follows:

Sitting

Easiest: Sitting on a chair or exercise ball

First Progression: Sitting on an exercise ball with arms crossed in front of chest and one knee in full extension.

Second Progression: Sitting on an exercise ball with arms crossed in front of chest and one knee in full extension and eyes closed.

Standing

Easiest: Standing on one foot.

First Progression: Standing on one foot while slowly raising and lowering arms.

Second Progression: Standing on one foot on a foam pad while slowly raising and lowering arms.

Dynamic

Easiest: Heel-to-toe walking in a straight line.

First Progression: Walking lunges in a straight line.

Second Progression: Walking lunges in a straight line while receiving and throwing a tennis ball.

The volume of balance exercises can be done for time, as in flexibility training. For example, holding the single-leg balance exercise for three sets of 30 s. It can also be prescribed in repetitions, as with resistance training. For example, completing walking lunges with tennis ball receiving for a total of two sets of 12 repetitions.

FUNCTIONAL FITNESS TESTING

Name _____ Sex _____ Date _____

Age _____   Height _____cm

Functional Fitness Testing Battery

30-second chair stand	Repetitions _____	Rating _____
30-second arm curl	Repetitions _____	Rating _____
Chair sit-and-reach	L _____ in. _____ in. R _____ in. _____ in.	Rating _____ Rating _____
Back scratch	LoR _____ in. _____ in. RoL _____ in. _____ in.	Rating _____ Rating _____
8-foot up-and-go	Trial 1 _____ s Trial 2 _____ s Trial 3 _____ s Avg: _____ s	Rating _____
6-minute walk test	Distance _____ yd	Rating _____
2-minute step test	Steps _____	Rating _____

Balance Error Scoring System

Stance	Firm surface error tally	Foam surface error tally	Total errors
Narrow double leg			
Single leg			
Tandem			

Total Error Score _____   Rating _____

Gait Speed

Trial 1	Trial 2	Trial 3	Average time	Average speed

Rating _____

From J. Janot and N. Beltz, *Laboratory Assessment and Exercise Prescription* (Champaign, IL: Human Kinetics, 2023).

BALANCE EXERCISE PROGRESSION PLANNING SHEET

Complete the balance exercise progression planning sheet by creating a 3 d·wk^{-1} balance program for 10 wk. Start with basic exercises for sitting, standing, and walking. Slowly progress each exercise through basic, moderate, and advanced tasks.

Day 1

Weeks 1-3 Basic		Weeks 4-7 Moderate		Weeks 8-10 Advanced	
Mode	Exercise	Mode	Exercise	Mode	Exercise
Sitting		Sitting		Sitting	
Standing		Standing		Standing	
Walking		Walking		Walking	

Day 2

Weeks 1-3 Basic		Weeks 4-7 Moderate		Weeks 8-10 Advanced	
Mode	Exercise	Mode	Exercise	Mode	Exercise
Sitting		Sitting		Sitting	
Standing		Standing		Standing	
Walking		Walking		Walking	

Day 3

Weeks 1-3 Basic		Weeks 4-7 Moderate		Weeks 8-10 Advanced	
Mode	Exercise	Mode	Exercise	Mode	Exercise
Sitting		Sitting		Sitting	
Standing		Standing		Standing	
Walking		Walking		Walking	

From J. Janot and N. Beltz, *Laboratory Assessment and Exercise Prescription* (Champaign, IL: Human Kinetics, 2023).

APPENDIX A: COMMON MEDICATIONS AND EXERCISE CONSIDERATIONS

This appendix includes information on common medications that you may come across when working with and designing exercise programs for apparently healthy individuals as well as those from clinical populations. The appendix is organized by main section headings of various conditions that these medications are designed to treat. Under these section headings, the medications are grouped within each table by their specific class and information, such as common names (generic and brand), action, therapeutic effect, and considerations for exercise. This appendix is not designed to be a comprehensive listing of medications that you may see when working with clients. Therefore, be prepared to do some research and discovery on your own if a different medication is brought to your attention. Some online resources you may find useful include Medscape and the U.S. Food and Drug Association (FDA), or hardcopy resources such as the *Nursing Drug Handbook*.

Common Medications for Managing Cardiac Conditions

This section includes a listing of medications under their individual therapeutic classes. These classes include antihypertensive agents, cholesterol- and triglyceride-lowering agents, antithrombotic (antiplatelet and anticoagulant) agents, and other cardiac agents.

Antihypertensive Agents

The antihypertensive medications in this section are primarily prescribed to control blood pressure (BP) at rest and during other activities such as exercise. Most of these medications do not alter the systolic BP and heart rate (HR) response to exercise. The exceptions in this case are the beta-blockers, which is why this class of medication is often associated with exercise intolerance and issues like hypotension during or following exercise. Clients taking beta-blockers need to be monitored closely and have their BP checked regularly during all phases of the exercise session. All antihypertensive agents will lower maximal BP responses during incremental exercise mostly due to decreased BP at rest.

CATEGORY	MEDICATION INFORMATION
Class of medication	Beta-blockers (β-blockers)
Common names (generic/brand)	Atenolol/Tenormin (β-1 cardioselective) Metoprolol/Lopressor (β-1 cardioselective) Acebutolol/Sectral (Class II antiarrhythmic) Pindolol/Visken (Class II antiarrhythmic) Timolol/Blocadren (Non-cardioselective) Nadolol/Corgard (Non-cardioselective) Propranolol/Inderal (Non-cardioselective) Carvedilol/Coreg (α-1 and non-cardioselective β-blocker)
Action	In general, β-blockers compete with catecholamines for β-adrenergic receptors. Cardioselective β-blockers block β-1 receptors only and have little effect on the bronchioles of the lung. Non-cardioselective β-blockers block both β-1 and β-2 receptors. Carvedilol blocks all β-adrenergic and α-1 receptors. Because of this action, β-blockers are referred to as negative inotropic (contractility) and chronotropic (HR) agents and can be used to control cardiac arrhythmias.
Therapeutic effect	The main effects of β-blockers are decreased HR and BP at rest and exercise. Collectively, this leads to lower myocardial work and oxygen demand. They are indicated for individuals with a history of hypertension, myocardial infarction, and angina pectoris (chest pain) to either prevent or manage myocardial ischemia and angina. Carvedilol is prescribed for clients with heart failure to improve cardiac function and disease management. β-blockers will also be prescribed to manage migraine headaches in some clients.
Considerations for exercise	Due to the therapeutic effects of β-blockers, clients should be monitored closely, and BP and HR should be checked prior to and during any exercise activity. Any symptoms of dizziness, lightheadedness, or other related issues should be reported. Cool-down activity following exercise should be prolonged (up to 10 min) to allow for appropriate adjustment of BP back to resting levels and to avoid a hypotensive response. β-blockers can increase exercise capacity during physical activity in those who have a history of angina and heart failure. Use of traditional HR prescription methods and prediction of maximal HR cannot be done in these clients. Thus, alternatives such as rating of perceived exertion (RPE), talk test, and other non-HR methods need to play a greater role in prescribing exercise intensity. To use HR prescription methods, a maximal exercise test needs to be done to determine maximal HR while taking β-blockers.

CATEGORY	MEDICATION INFORMATION
Class of medication	Alpha-1 blockers (α-1 blockers)
Common names (generic/brand)	Doxazosin/Cardura Prazosin/Minipress Terazosin/Hytrin
Action	α-1 blockers compete with norepinephrine at the postsynaptic α-1 adrenergic receptor.
Therapeutic effect	Administration causes arterial and venous dilation and a subsequent decrease in BP during rest and exercise due to an overall drop in total peripheral resistance (afterload). They are indicated for individuals with a history of hypertension, and are given to manage benign prostatic hyperplasia.
Considerations for exercise	Due to the therapeutic effects of α-1 blockers, BP should be monitored closely and checked prior to and during any exercise activity. Any symptoms of dizziness, lightheadedness, or other related issues should be reported. Cool-down activity for adjustment of BP following exercise should be completed to avoid a hypotensive response. Traditional HR prescription methods and prediction of maximal HR can be used for exercise programming.

CATEGORY	MEDICATION INFORMATION
Class of medication	Angiotensin-converting enzyme inhibitors (ACEIs)
Common names (generic/brand)	Benazepril/Lotensin Captopril/Capoten Enalapril/Vasotec Lisinopril/Prinivil or Zestril Quinapril/Accupril Ramapril/Altace
Action	Inhibits angiotensin-converting enzyme (ACE) from transforming the hormone angiotensin I into the vasoconstrictor factor angiotensin II. ACE inhibition also decreases release of the hormone aldosterone, which leads to greater Na^+ and water excretion by the kidney, and decreased plasma volume.
Therapeutic effect	ACEIs cause peripheral vasodilation and a subsequent decrease in BP during rest and exercise due to an overall drop in total peripheral resistance (afterload) and preload. ACEIs are indicated for individuals with a history of hypertension and myocardial infarction, and are first-line medications given to manage clients in heart failure with reduced ejection fraction (HFrEF). They also elicit a protective effect over the kidney by promoting vasodilation of renal arteries.
Considerations for exercise	Due to the therapeutic effects of ACEIs, BP should be checked prior to and during any exercise activity. Any symptoms of dizziness, lightheadedness, or other related issues should be reported. Cool-down activities should be prescribed to monitor for adjustment of BP back to resting levels following exercise. ACEIs increase exercise capacity and tolerance during physical activity in those who have a history of HFrEF. With little effect on resting and exercise HR, traditional HR prescription methods and prediction of maximal HR can be used for these clients during exercise.

CATEGORY	MEDICATION INFORMATION
Class of medication	Angiotensin II receptor blockers (ARBs)
Common names (generic/brand)	Candesartan/Atacand Losartan/Cozaar Valsartan/Diovan
Action	ARBs block receptors targeted by angiotensin II. This inhibition will decrease BP through vasodilation and decreased plasma volume. ARBs further inhibit the release of the hormone aldosterone, which leads to greater Na^+ and water excretion by the kidney, and decreased plasma volume.
Therapeutic effect	ARBs cause peripheral vasodilation and a subsequent decrease in BP during rest and exercise due to an overall drop in total peripheral resistance (afterload) and preload. ARBs are indicated for individuals with a history of hypertension and myocardial infarction, and are given to manage clients in heart failure.
Considerations for exercise	BP should be closely monitored prior to and during any exercise activity. Dizziness, lightheadedness, or other related symptoms should be reported immediately. Prescribing cool-down activities following exercise should help with appropriate adjustment of BP back to resting levels, and decrease the risk of a hypotensive response. ARBs may increase exercise capacity and tolerance during physical activity in those who have a history of HFrEF. Because there is little effect on HR at rest and during exercise, traditional HR prescription methods and prediction of maximal HR can be used.

CATEGORY	MEDICATION INFORMATION
Class of medication	Calcium channel blockers (CCBs)
Common names (generic/brand)	Verapamil/Calan (Class IV antiarrhythmic) Diltiazem/Cardizem (Class IV antiarrhythmic) Amlodipine/Norvasc Felodipine/Plendil Nicardipine/Cardene Nifedipine/Procardia
Action	CCBs inhibit influx of extracellular calcium ions into both myocardial and vascular smooth muscle cells, resulting in decreased cardiac contractility and vascular smooth muscle contraction (vasoconstriction). This will lead to vasodilation in both coronary and systemic arteries and lowering total peripheral resistance (afterload). Because of this action, CCBs are considered to be negative inotropic (contractility) agents. Some CCBs slow inward calcium current responsible for depolarization, acting principally at the atrioventricular (AV) node (diltiazem) with some effects at both the sinoatrial (SA) and AV node (verapamil). This slows impulse conduction down through the cardiac conduction system.
Therapeutic effect	The main effects of CCBs are decreased BP at rest and exercise, with a small impact on resting HR. Collectively, this leads to lower myocardial work and oxygen demand. CCBs are indicated for individuals with a history of hypertension, coronary artery disease, and angina pectoris (chest pain) to either prevent or manage myocardial ischemia and angina. Certain types of CCBs are prescribed to control supraventricular arrhythmias (e.g., atrial fibrillation or flutter, and supraventricular tachycardia) plus coronary vasospasm or pulmonary hypertension in some cases.
Considerations for exercise	Due to the therapeutic effects of CCBs, clients should be monitored closely for exercise-induced hypotension. Any adverse symptoms during or following exercise should be reported. Cool-down activity following exercise should be prolonged (up to 10 min) to allow for appropriate adjustment of BP back to resting levels. CCBs can increase exercise capacity in those who have a history of angina. Traditional HR prescription methods and prediction of maximal HR can be utilized during exercise.

CATEGORY	MEDICATION INFORMATION
Class of medication	**Diuretics**
Common names (generic/brand)	Furosemide/Lasix (loop diuretic) Bumetanide/Bumex (loop diuretic) Chlorthalidone/Chlorthalid (thiazide diuretic) Hydrochlorothiazide/Esidrix or HydroDIURIL (thiazide diuretic) Indapamide (thiazide diuretic) Spironolactone/Aldactone (aldosterone receptor antagonist)
Action	Loop diuretics: Work at the ascending loop of Henle, inhibiting the reabsorption of Na^+, Cl^-, and water, increasing renal urine output. Thiazide diuretics: Inhibit the reabsorption of Na^+ in the distal convoluted tubule of the nephron by blocking the Na^+ and Cl^- channels, increasing renal urine output. Aldosterone receptor antagonists: Compete with aldosterone for receptors in the distal convoluted tubule and collecting duct. Blockade of aldosterone inhibits Na^+, Cl^-, and water reabsorption in the renal tubules and increases retention of K^+.
Therapeutic effect	The main effects of diuretics are decreased BP through reduction in plasma volume and preload. Collectively, this leads to lower myocardial work and oxygen demand. Resting HR could be elevated due to potential dehydration from diuretic therapy. These medications are indicated for individuals with a history of hypertension, edema secondary to chronic disease (e.g., heart failure, liver disease, kidney disease, pulmonary disease), and heart failure (both HFrEF and heart failure with preserved ejection fraction). Loop diuretics are usually prescribed to manage acute exacerbations of various chronic conditions or during emergency situations, whereas thiazide and aldosterone antagonist diuretics are prescribed for long-term therapeutic reasons such as managing hypertension or heart failure.
Considerations for exercise	Due to the therapeutic effects of diuretics, clients should be monitored closely for any signs or symptoms of dizziness, lightheadedness, rapid weight changes, recent or new arrhythmias, or other unfavorable issues. To prevent hypotensive responses following exercise, cool-down activity should be stressed. Use of traditional HR prescription methods and prediction of maximal HR is appropriate for clients taking diuretics.

Cholesterol- and Triglyceride-Lowering Agents

CATEGORY	MEDICATION INFORMATION
Class of medication	HMG-CoA reductase inhibitors (statins)
Common names (generic/brand)	Atorvastatin/Lipitor Fluvastatin/Lescol Pravastatin/Pravachol Rosuvastatin/Crestor Simvastatin/Zocor Lovastatin/Mevacor
Action	Inhibits HMG-CoA reductase, which is the rate-limiting enzyme in the process of cholesterol synthesis in the liver.
Therapeutic effect	Indicated for individuals at risk for or with a history of hypercholesterolemia, to lower low-density lipoprotein (LDL) cholesterol. Possibly improves nitric oxide production, leading to improved vascular health and decreased clot formation.
Considerations for exercise	Because of statin use, clients may experience a side effect of prolonged and significant muscle soreness, which could be exacerbated by exercise. If clients who are taking this medication complain of muscle soreness lasting more than 72 h post-exercise, they should contact their healthcare provider. It would be appropriate to lower exercise intensity or discontinue exercise until this is resolved. No other significant effect on the exercise response is present.

CATEGORY	MEDICATION INFORMATION
Class of medication	Fibric acid agents
Common names (generic/brand)	Gemfibrozil/Lopid Fenofibrate/Tricor
Action	Increase the activity of lipoprotein lipase (LPL) leading to greater breakdown of very low-density lipoproteins (VLDL) and uptake and oxidation of free-fatty acids (FFA).
Therapeutic effect	Indicated for individuals at risk for or with a history of hypertriglyceridemia and hypercholesterolemia, to lower triglycerides and VLDL.
Considerations for exercise	These medications are commonly paired with statins; thus, a side effect of increased muscle soreness is a possibility, which could be exacerbated by exercise. This medication alone may trigger generalized muscle soreness. If individuals complain of prolonged muscle soreness, it is important to contact their healthcare provider. Lower the exercise intensity or discontinue exercise until this is resolved. Otherwise, no significant effect on the exercise response is noted with these medications.

CATEGORY	MEDICATION INFORMATION
Class of medication	Bile acid sequestrants
Common names (generic/brand)	Colesevelam/WelChol Cholestyramine/Questran Colestipol/Colestid
Action	Binds with bile acids not absorbed through the intestine and inhibits liver reuptake of intestinal bile salts. This leads to excretion of bile salt-bound LDL cholesterol through the GI tract.
Therapeutic effect	Reduction of serum cholesterol and indicated for individuals at risk for or with a history of hyperlipidemia.
Considerations for exercise	No significant effect on the exercise response.

CATEGORY	MEDICATION INFORMATION
Class of medication	Other lipid-lowering agents
Common names (generic/brand)	Niacin/Niaspan Omega-3 fatty acids/Lovaza Rosuvastatin + Ezetimibe/Roszet Simvastatin + Ezetimibe/Vytorin
Action	Niacin: Inhibition of VLDL synthesis in the liver. Omega-3 fatty acid supplementation: May decrease liver triglyceride synthesis, increase LPL activity to decrease blood triglycerides, and lower proinflammatory marker concentration in the blood (anti-inflammatory effect). Roszet and Vytorin: Combination effect inhibiting both the activity of the rate-limiting enzyme HMG-CoA reductase in the liver (statin) and the intestinal absorption of cholesterol (Ezetimibe).
Therapeutic effect	Lowers serum cholesterol and triglycerides. Indicated for individuals at risk for or with a history of hyperlipidemia (niacin), hypercholesterolemia (Roszet and Vytorin), and hypertriglyceridemia (omega-3 fatty acids or niacin).
Considerations for exercise	Because of statin use with Roszet and Vytorin, there may be a side effect of muscle soreness that can be exacerbated by exercise. If clients who are taking this medication complain of prolonged muscle soreness, it is important to contact their healthcare provider. In these cases, lower exercise intensity or discontinue exercise until resolved. No other significant effect on the exercise response is present.

Antithrombotic Agents

CATEGORY	MEDICATION INFORMATION
Class of medication	Anticoagulants
Common names (generic/brand)	Warfarin/Coumadin Rivaroxaban/Xarelto Apixaban/Eliquis
Action	Coumadin: Disrupts the formation of vitamin K–dependent clotting factors in the liver. Xarelto and Eliquis: Inhibit platelet activation and the blood coagulation cascade through decreasing the activity of factor Xa.
Therapeutic effect	Decrease the formation of clots linked to thromboembolism in both arteries and veins eliciting tissue ischemia. Indicated for individuals at risk for or with a history of stroke, cardiac valve disease, myocardial infarction, atrial fibrillation, and deep vein thrombosis.
Considerations for exercise	No effect on the exercise response.

CATEGORY	MEDICATION INFORMATION
Class of medication	Antiplatelets
Common names (generic/brand)	Acetylsalicylic acid (ASA)/Aspirin (various manufacturers) Clopidogrel/Plavix
Action	Aspirin: Decreases synthesis of prostaglandins through inhibition of cyclooxygenase-1, leading to less platelet aggregation and clot formation. Plavix: Inhibits the binding of adenosine diphosphate to its target receptor, inhibiting the platelet activation and aggregation pathway.
Therapeutic effect	Depending on the specific medication, indicated for individuals at risk for or with a history of ischemic stroke and transient ischemic attack, acute coronary syndrome, atherosclerotic cardiovascular disease, myocardial infarction, and peripheral artery disease (PAD).
Considerations for exercise	Plavix could increase exercise capacity and limit the effects of intermittent claudication in individuals with PAD. Otherwise, there are no significant effects on the exercise response.

Other Cardiac Agents

CATEGORY	MEDICATION INFORMATION
Class of medication	Angiotensin receptor-neprilysin inhibitors (ARNIs)
Common names (generic/brand)	Sacubitril + Valsartan/Entresto
Action	Sacubitril: Inhibits the enzyme neprilysin, which degrades substances called natriuretic peptides. Inhibition results in increased urine output, decreased plasma volume, decreased BP, and less overall stress on the heart. Valsartan: Blocks receptors targeted by angiotensin II. This inhibition will decrease BP through vasodilation and decreased plasma volume.
Therapeutic effect	ARNIs elicit a decrease in BP at rest and during exercise. They are indicated for individuals in HFrEF to reduce risk of mortality and hospitalization.
Considerations for exercise	Increase exercise capacity and tolerance in individuals with HFrEF.

CATEGORY	MEDICATION INFORMATION
Class of medication	Antiarrhythmic agents
Common names (generic/brand)	Digoxin/Lanoxin Disopyramide/Norpace (Class Ia medication) Propafenone/Rythmol (Class Ic medication) Amiodarone/Cordarone (Class III medication) Sotalol/Betapace (Class III medication)
Action	Digoxin: Inhibits the Na^+/K^+ pump, which increases intracellular calcium in the myocardium and cardiac conduction system. Norpace and Rythmol: Slow conduction velocity of the action potential in cardiac tissue by altering the transport of Na^+ and other ions across the cardiac myocyte membranes. This will also prolong impulse refractoriness and decrease automaticity to provide cardioprotection from provoked arrythmias. Amiodarone and sotalol: Inhibit adrenergic (sympathetic nervous system) stimulation, slow the action potential down through the cardiac conduction system, decrease AV node impulse conduction, and decrease SA node firing rate.

CATEGORY	MEDICATION INFORMATION
Class of medication	Antiarrhythmic agents
Therapeutic effect	Digoxin: Increased intracellular calcium leads to a positive inotropic effect on the myocardium (increased force of contraction). Reduced conduction velocity in the cardiac conduction system has an antiarrhythmic effect and negative chronotropic effect in individuals with heart failure. Norpace and Rythmol: Indicated for individuals at risk for serious ventricular tachyarrhythmias (e.g., ventricular tachycardia). Rythmol can also be used to control a variety of arrhythmias (e.g., atrial fibrillation/flutter, supraventricular tachycardia). Amiodarone and sotalol: Indicated for individuals at risk for serious ventricular tachyarrhythmias (e.g., ventricular tachycardia and ventricular fibrillation). Sotalol can be used to control atrial fibrillation/flutter.
Considerations for exercise	Digoxin increases exercise capacity and tolerance for individuals with HFrEF and atrial fibrillation. It is also used to control atrial fibrillation, and can decrease HR at rest and during exercise. The normal ECG can be affected by this medication; thus, graded exercise tests with ECG only are not recommended. There is no significant effect on BP. Amiodarone and propafenone may elicit a lower HR during rest and exercise, and amiodarone may increase BP during exercise. Exercise capacity can be increased in these clients due to overall cardiac rate and rhythm control. Traditional HR prescription methods can be used in these clients.

CATEGORY	MEDICATION INFORMATION
Class of medication	Nitrates (vasodilator agents)
Common names (generic/brand)	Nitroglycerin/Nitro-Dur, Nitrolingual, Nitrostat, Nitro-Bid Isosorbide mononitrate/Imdur Isosorbide dinitrate/Dilatrate or Isordil
Action	Enters vascular smooth muscle and is converted to nitric oxide leading to overall vasodilation. This medication can be delivered as follows: pill form under tongue, spray on or under tongue, skin patch, short or long extended-release in pill form, or skin ointment.
Therapeutic effect	Decreases myocardial work and oxygen demand through reduction of mostly cardiac preload and some afterload. Indicated for individuals with a history of angina pectoris (chest pain) to either prevent or manage myocardial ischemia and angina. This medication is associated with an acute lowering of BP, increased HR, and occasional paradoxical bradycardia (slow heart rate following administration).
Considerations for exercise	Due to its therapeutic effects, clients should be monitored closely after taking this medication: BP and HR should be checked prior to any exercise activity, and symptoms of dizziness, lightheadedness, or other related issues should be noted. This medication can also be taken prophylactically prior to exercise to reduce the incidence of angina and increase exercise capacity in those who have a history of angina. HR prescription methods can be used in these clients.

Common Medications for Managing Metabolic (Diabetes) Conditions

This section includes a listing of medications used to manage diabetes mellitus. In general, the oral hypoglycemic medications are utilized by individuals with type 2 diabetes to manage their blood glucose and prevent chronic conditions from occurring, which are linked to poor blood glucose control. Insulin is the primary agent used to control blood glucose for individuals with type 1 diabetes; however, individuals with longstanding type 2 diabetes could be taking insulin to manage their blood glucose as well. It is important for any individual with diabetes to check their blood sugars prior to, during (if needed), and after exercise and adjust exercise time and nutrition accordingly.

Oral Hypoglycemic Agents

CATEGORY	MEDICATION INFORMATION
Class of medication	Biguanides
Common names (generic/brand)	Metformin/Glucophage
Action	Lowers liver glucose production, improves insulin sensitivity at the muscle, and decreases intestinal absorption of glucose.
Therapeutic effect	Lowers blood glucose.
Considerations for exercise	Risk of exercise-induced hypoglycemia is low when used alone and not in combination with other blood glucose–lowering medications.

CATEGORY	MEDICATION INFORMATION
Class of medication	Thiazolidinediones
Common names (generic/brand)	Pioglitazone/Actos Rosiglitazone/Avandia
Action	Increases insulin sensitivity and glucose uptake at a variety of tissues (e.g., muscle, adipose), decreases liver glucose production through the gluconeogenesis pathway, and decreases plasma FFA concentrations.
Therapeutic effect	Lowers blood glucose.
Considerations for exercise	Risk of exercise-induced hypoglycemia is low when used alone and not in combination with other blood glucose–lowering medications, especially insulin.

CATEGORY	MEDICATION INFORMATION
Class of medication	Dipeptidyl peptidase-4 (DPP-4) inhibitors
Common names (generic/brand)	Sitagliptin/Januvia Saxagliptin/Onglyza Linagliptin/Tradjenta
Action	Increases the activity of the incretin hormone glucagon-like peptide-1 (GLP-1) by inhibiting the DPP-4 enzyme. This leads to increased insulin and decreased glucagon secretion from the pancreas and less glucose production from the liver.
Therapeutic effect	Lowers blood glucose.
Considerations for exercise	Risk of exercise-induced hypoglycemia is low when used alone and not in combination with other blood glucose–lowering medications, such as sulfonylureas and meglitinides.

CATEGORY	MEDICATION INFORMATION
Class of medication	Sodium-glucose cotransporter-2 (SGLT$_2$) inhibitors
Common names (generic/brand)	Canagliflozin/Invokana Dapagliflozin/Farxiga Empagliflozin/Jardiance
Action	Decreases glucose reabsorption into the blood in the proximal convoluted tubule area of the nephron. More glucose remains in the nephron and is excreted through the urine.
Therapeutic effect	Lowers blood glucose; decreases mortality rate in heart failure.
Considerations for exercise	Risk of exercise-induced hypoglycemia is low when used alone and not in combination with other blood glucose–lowering medications, especially insulin.

CATEGORY	MEDICATION INFORMATION	
Class of medication	Sulfonylureas	Meglitinides
Common names (generic/brand)	Glipizide/Glucotrol Glyburide/Micronase	Repaglinide/Prandin Nateglinide/Starlix
Action	Stimulates the beta cells of the pancreas to increase insulin production. Sulfonylureas have a longer therapeutic action time compared to meglitinides.	
Therapeutic effect	Lowers blood glucose.	
Considerations for exercise	There is increased risk of exercise-induced hypoglycemia with these oral medications because they are related more directly to insulin secretion from the pancreas. Dosing may need to be adjusted by a healthcare provider on exercise days and blood glucose closely monitored pre- and post-exercise if hypoglycemia has been an issue.	

CATEGORY	MEDICATION INFORMATION
Class of medication	Incretin mimetics
Common names (generic/brand)	Liraglutide/Victoza Dulaglutide/Trulicity
Action	Mimics the action of the incretin hormone GLP-1 to increase insulin and decrease glucagon secretion from the pancreas. This will lead to less glucose released from the liver.
Therapeutic effect	Lowers blood glucose.
Considerations for exercise	Risk of exercise-induced hypoglycemia is low when used alone and not in combination with other blood glucose–lowering medications, especially insulin or sulfonylureas.

Insulin

CATEGORY	MEDICATION INFORMATION
Class of medication	Insulin
Common names (generic/brand)	Long-acting insulin: Glargine/Lantus Intermediate-acting insulin: NPH/Humulin R or Novolin R Regular or short-acting insulin: Regular/Humulin N or Novolin N Rapid-acting insulin: Lispro/Humalog or Aspart/NovoLog
Action	Increases the uptake of blood glucose into peripheral cells through insulin-mediated binding mechanisms, increased FFA uptake in adipose tissue, increased protein synthesis, and decreased glucose release by the liver.
Therapeutic effect	Lowers blood glucose.
Considerations for exercise	Risk of exercise-induced hypoglycemia is highest in individuals taking insulin. Three main points to understand regarding insulin is how fast it starts working (onset), when action peaks in the blood (peak time), and how long it stays in the blood at effective levels (duration of action). This information can help prevent hypoglycemic episodes. Managing insulin dosage timing with meals and exercise, and specific to the current level of blood glucose, can be a complicated undertaking. Therefore, it is critical that the exercise professional is communicating with the client and the client's healthcare provider regarding the exercise plan.

ADDITIONAL INFORMATION ON INSULIN

Type of insulin	Onset	Peak time	Duration of action
Long-acting insulin	2-3 h	None	24 h
Intermediate-acting insulin	2-4 h	4-12 h	12-18 h
Regular (short-acting) insulin	30 min	2-3 h	3-6 h
Rapid-acting insulin	15 min	1-2 h	2-4 h

Common Medications for Managing Pulmonary Conditions

This section includes a listing of medications used to manage pulmonary disease, specifically chronic asthma (bronchoconstriction) and chronic obstructive pulmonary disease (COPD). These medications are commonly seen in individuals across the lifespan, especially in those looking to manage the effects of asthma and other chronic conditions so that they can engage in physical activity and activities of daily living more safely and effectively.

CATEGORY	MEDICATION INFORMATION
Class of medication	Inhaled corticosteroids
Common names (generic/brand)	Fluticasone/Flovent Budesonide/Pulmicort Beclomethasone/Qvar
Action	Anti-inflammatory agents that bind to glucocorticoid receptors in bronchiole smooth muscle inhibiting the action of cells (e.g., macrophages, neutrophils, mast cells, etc.) and mediators (e.g., histamine, cytokines, etc.) of the immune system.
Therapeutic effect	Utilized as maintenance therapy for COPD and asthma. Improves airflow and ventilation by decreasing airway obstruction caused by inflammation (i.e., increased mucous production, edema, bronchoconstriction, etc.).
Considerations for exercise	Improved exercise tolerance in individuals limited by acute or chronic bronchospasm and COPD. No cardiovascular effects during exercise.

CATEGORY	MEDICATION INFORMATION
Class of medication	Beta agonists (sympathomimetics)
Common names (generic/brand)	Short-acting beta agonist: Albuterol/Ventolin Long-acting beta agonist: Salmeterol/Serevent or formoterol/Foradil
Action	Relaxes bronchiole smooth muscle by stimulating β_2-receptors in the bronchioles.
Therapeutic effect	Decreases bronchospasm to lower resistance to airflow in the lung and improve ventilation. Albuterol works much faster over a short period of time to reverse bronchospasm quickly. Salmeterol works slower but over a longer period of time for better control and prevention of bronchospasm.
Considerations for exercise	Improved exercise tolerance in individuals limited by acute or chronic bronchospasm and COPD. Little to no effect on HR and BP.

CATEGORY	MEDICATION INFORMATION
Class of medication	Anticholinergics
Common names (generic/brand)	Short-acting muscarinic antagonist: Ipratropium/Atrovent Long-acting muscarinic antagonist: Tiotropium/Spiriva
Action	Inhibits the binding of acetylcholine to muscarinic receptors on bronchiole smooth muscle.
Therapeutic effect	Utilized as maintenance therapy for COPD and asthma. Decreases bronchoconstriction to lower resistance to airflow in the lung, decreases airway inflammation, decreases mucous production, and decreases airway hyperreactivity to agents that can trigger bronchospasm. These effects lead to less airway obstruction and improved ventilation. Spiriva is a longer-acting agent for longer-term control of COPD and asthma.
Considerations for exercise	Improved exercise tolerance in individuals limited by acute or chronic bronchoconstriction and COPD. Little to no effect on HR and BP.

CATEGORY	MEDICATION INFORMATION
Class of medication	Combination agents
Common names (generic/brand)	Salmeterol + Fluticasone/Advair Formoterol + Budesonide/Symbicort Ipratropium + Albuterol/Combivent
Action	Advair and Symbicort: Combination bronchodilator and corticosteroid to induce bronchodilation and decreased airway inflammation. Combivent: Combination bronchodilator and anticholinergic to induce bronchodilation, decrease airway inflammation, and decrease airway hyperreactivity.
Therapeutic effect	Decreases bronchospasm and chronic airway inflammation to lower resistance to airflow in the lung and improve ventilation. These agents are used to provide better control over chronic airway obstruction commonly observed in chronic asthma and COPD.
Considerations for exercise	Improved exercise tolerance in individuals limited by acute or chronic bronchospasm and COPD. Little to no effect on HR and BP. Could provoke tachyarrhythmias as a side effect.

Exercise testing performed in an exercise physiology laboratory is a generally safe procedure. The event rate of hospitalization or death during an exercise test is approximately 1 in every 10,000 tests (Liguori and ACSM 2021). Despite the low adverse incidence rate, personnel working in an exercise physiology laboratory setting should possess a minimum level of training. In addition, the emergency procedures of each laboratory should be visibly posted and reviewed by all staff on an annual basis. This textbook is designed for exercise physiology laboratories in an educational setting. Therefore, an on-site medical director is not always a realistic expectation. Minimizing risk should be considered in the educational setting. For example, community members may use the lab facility to have routine resting and exercise ECG testing. Resting ECG administration does not require an on-site medical director, but the results of the test should be sent to a medical director for interpretation. An exercise ECG test may require the presence of a medical director if the participant has known cardiovascular, pulmonary, renal, or metabolic disease. All staff within the lab should have a minimum of Basic Life Support (BLS) training that covers cardiopulmonary resuscitation (CPR) and automated external defibrillator (AED). Laboratories that work with individuals with known cardiovascular, pulmonary, renal, or metabolic disease may want additional training in Advanced Cardiovascular Life Support (ACLS) for the management of cardiac arrest, arrythmias, airway management, pharmacology, stroke, and resuscitation team communication skills.

All laboratory staff should know the following information:

- AED locations within the lab and building.
- Office phone locations. Personal cell phones may also be available for emergency use.
- Physical address of the building and laboratory room.

If a life-threatening situation has been established, the client is down, and there is any question about breathing, pulse, or consciousness, do the following:

1. Call for help from a coworker in the lab. Instruct the coworker to call 911 on their cell phone or the nearest office phone and to bring you the nearest AED. If you are calling from an office phone, be aware of any extensions that need to be dialed prior to calling an outside number.
2. Have a planned script explaining the nature of the emergency, the exact location of the emergency, and where the emergency medical service (EMS) team can expect to meet you.
3. Stay on the phone until the EMS disconnects.
4. Continue assessing the situation. If necessary, perform CPR in accordance with BLS or ACLS standards.
5. Upon arrival of the AED, follow BLS or ACLS standards for AED use.

If a non-life-threatening situation has been established, such as the client fell but is conscious, has orthopedic pain, abnormal fatigue, dizziness, nausea, and so forth, do the following:

1. Assess the severity of the situation.
2. Have the client sit or lie down if necessary.
3. Ask the client for any relevant symptoms. Are they new? Are they reoccurring? How severe are they?
4. Remain with the client until symptoms lessen.
5. If symptoms worsen or the client loses consciousness, follow the procedures for a life-threatening situation.

REFERENCES

Introduction

Albarqouni, L., T. Hoffmann, S. Straus, N.R. Olsen, T. Young, D. Ilic, T. Shaneyfelt, et al. 2018. "Core Competencies in Evidence-Based Practice for Health Professionals: Consensus Statement Based on a Systematic Review and Delphi Survey." *The Journal of the American Medical Association Network Open* 1:e180281.

American College of Sports Medicine (ACSM). 2021. *ACSM's Guidelines for Exercise Testing and Prescription*. 11th ed. Baltimore: Lippincott Williams & Wilkins.

Hall, K., T. Gibbie, and D.I. Lubman. 2012. "Motivational Interviewing Techniques: Facilitating Behaviour Change in the General Practice Setting." *Australian Family Physician* 41:660-667.

Jackson, A. and M. Pollock. 1985. "Practical Assessment of Body Composition." *The Physician and Sportsmedicine* 13:76-90.

Sackett, D.L., W.M.C. Rosenberg, J.A. Muir Gray, R.B. Haynes, and W.S. Richardson. 1996. "Evidence Based Medicine: What It Is and What It Isn't." *British Medical Journal* 312:71-72.

Lab 1

American College of Sports Medicine (ACSM). 2021. *ACSM's Guidelines for Exercise Testing And Prescription*. 11th ed. Baltimore: Lippincott Williams & Wilkins.

Benjamin, E., P. Munter, A. Alonso, M. Bittencourt, C. Callaway, A. Carson, A. Chamberlain, et al. 2019. "Heart Disease and Stroke Statistics–2019 Update." *Circulation* 139:e56-e528.

Brannick, B. and S. Dagogo-Jack. 2018. "Prediabetes and Cardiovascular Disease: Pathophysiology and Interventions for Prevention and Risk Reduction." *Endocrinology and Metabolism Clinics of North America* 47:33-50.

Centers for Disease Control and Prevention (CDC). 2020. "Assessing Your Weight." Last modified September 17, 2020. www.cdc.gov/healthyweight/assessing/index.html.

Grundy, S., N. Stone, A. Bailey, C. Beam, K. Birtcher, R. Blumenthal, L. Braun, et al. 2019. "2018 ACC/AHA/AAPA/ABC/ACPM/AGS/APhA/ASH/ASPC/NMA/PCNA Guideline on the Management of Blood Cholesterol." *Circulation* 139:e1082-31143.

Lloyd-Jones, D., B. Nam, R. D-Agostino, D. Levy, J. Murabito, T. Wang, P. Wilson, and C. O'Donnell. 2004. "Parental Cardiovascular Disease as a Risk Factor for Cardiovascular Disease in Middle-aged Adults." *Journal of the American Medical Association* 291:2204-2211.

Rodgers, J., J. Jones, S. Bolleddu, S. Vanthenapalli, L. Rodgers, K. Shah, K. Karia, and S. Panguluri. 2019. "Cardiovascular Risks Associated With Gender and Aging." *Journal of Cardiovascular Development and Disease* 6:1-18.

U.S. Department of Health and Human Services (USDHHS). 1998. "Clinical Guidelines on the Identification, Evaluation and Treatment of Overweight and Obesity in Adults." *NIH Publication No. 98-4083*: 1-228.

Whelton, P., R. Carey, W. Aronow, D. Casey, K. Collins, C. Dennison Himmelfarb, S. DePalma, et al. 2018. "2017 ACC/AHA/AAPA/ABC/ACPM/AGS/APhA/ASH/ASPC/NMA/PCNA Guideline for the Prevention, Detection, Evaluation, and Management of High Blood Pressure in Adults. A Report of the American College of Cardiology/American Heart Association Task Force on Clinical Practice Guidelines." *Journal of the American College of Cardiology* 71:e127-e248.

World Health Organization (WHO). 2020a. "Tobacco." Last modified May 17, 2020. www.who.int/news-room/factsheets/detail/tobacco.

World Health Organization (WHO). 2020b. "Physical Activity." Last modified November 26, 2020. www.who.int/news-room/fact-sheets/detail/physical-activity.

Lab 2

Augustsson, S.R., E. Bersås, E. Magnusson Thomas, M. Sahlberg, J. Augustsson, and U. Svantesson. 2009. "Gender Differences and Reliability of Selected Physical Performance Tests in Young Women and Men." *Advances in Physiotherapy* 11(2):64-70. https://doi.org/10.1080/14038190801999679.

Binkley, N., D. Krueger, and B. Buehring. 2013. "What's in a Name Revisited: Should Osteoporosis and Sarcopenia Be Considered Components of 'Dysmobility Syndrome?'" *Osteoporosis International* 24(12):2955-2959. https://doi.org/10.1007/s00198-013-2427-1.

Blacher, J., J.A. Staessen, X. Girerd, J. Gasowski, L. Thijs, L. Liu, J.G. Wang, et al. 2000. "Pulse Pressure Not Mean Pressure Determines Cardiovascular Risk in Older Hypertensive Patients." *Archives of Internal Medicine* 160(8):1085-1089. https://doi.org/10.1001/archinte.160.8.1085.

Bohannon, R.W., S.R. Magasi, D.J. Bubela, Y.-C. Wang, and R.C. Gershon. 2012. "Grip and Knee Extension Muscle Strength Reflect a Common Construct among Adults." *Muscle & Nerve* 46(4):555-558. https://doi.org/10.1002/mus.23350.

Bray, G.A. and D.S. Gray. 1988. "Obesity. Part I—Pathogenesis." *The Western Journal of Medicine* 149(4):429-441.

Cameron, A.J., H. Romaniuk, L. Orellana, J. Dallongeville, A.J. Dobson, W. Drygas, M. Ferrario, et al. 2020. "Combined Influence of Waist and Hip Circumference on Risk of Death in a Large Cohort of European and Australian Adults." *Journal of the American Heart Association* 9(13):e015189. https://doi.org/10.1161/jaha.119.015189.

Canadian Society for Exercise Physiology. 2019. *Canadian Society For Exercise Physiology-Physical Activity Training For Health (CSEP-PATH)*. 2nd ed. Ottawa, ON, Canada: Canadian Society for Exercise Physiology.

DeMers, D. and D. Wachs. 2020. "Physiology, Mean Arterial Pressure." *StatPearls [Internet]*.

Deurenberg, P., J.A. Weststrate, and J.C. Seidell. 1991. "Body Mass Index as a Measure of Body Fatness: Age- and Sex-Specific Prediction Formulas." *British Journal of Nutrition* 65(2):105-114. https://doi.org/10.1079/bjn19910073.

Domanski, M.J., B.R. Davis, M.A. Pfeffer, M. Kastantin, and G.F. Mitchell. 1999. "Isolated Systolic Hypertension: Prognostic Information Provided by Pulse Pressure." *Hypertension* 34(3):375-380. https://doi.org/10.1161/01.hyp.34.3.375.

Emilio, E.J.M.L, F. Hita-Contreras, P.M. Jiménez-Lara, P. Latorre-Román, and A. Martínez-Amat. 2014. "The Association of Flexibility, Balance, and Lumbar Strength with Balance Ability: Risk of Falls in Older Adults." *Journal of Sports Science & Medicine* 13(2):349-357.

Fei, Y. 2020. "Understanding the Association between Mean Arterial Pressure and Mortality in Young Adults." *Postgraduate Medical Journal* 96(1138):453-454. https://doi.org/10.1136/postgradmedj-2020-137751.

Fleck, S.J. and W.J. Kraemer. 2004. *Designing Resistance Training Programs*. 3rd ed. Champaign, IL: Human Kinetics.

Franklin, S.S., L. Thijs, T.W. Hansen, E. O'Brien, and J.A. Staessen. 2018. "White-Coat Hypertension." *Hypertension* 62(6):982-987. https://doi.org/10.1161/hypertensionaha.113.01275.

Gibson, A.L., D.R. Wagner, and V.H. Heyward. 2019. *Advanced Fitness Assessment and Exercise Prescription*. 11th ed. Champaign, IL: Human Kinetics.

Guerra, R.S. and T.F. Amaral. 2009. "Comparison of Hand Dynamometers in Elderly People." *The Journal of Nutrition, Health & Aging* 13(10):907-912.

Hogrel, J.-Y. 2015. "Grip Strength Measured by High Precision Dynamometry in Healthy Subjects from 5 to 80 Years." *BMC Musculoskeletal Disorders* 16(1):139. https://doi.org/10.1186/s12891-015-0612-4.

Iwamoto, J., H. Suzuki, K. Tanaka, T. Kumakubo, H. Hirabayashi, Y. Miyazaki, Y. Sato, et al. 2009. "Preventative Effect of Exercise against Falls in the Elderly: A Randomized Controlled Trial." *Osteoporosis International* 20(7):1233-1240. https://doi.org/10.1007/s00198-008-0794-9.

Iwata, M., A. Yamamoto, S. Matsuo, G. Hatano, M. Miyazaki, T. Fukaya, M. Fujiwara, et al. 2019. "Dynamic Stretching Has Sustained Effects on Range of Motion and Passive Stiffness of the Hamstring Muscles." *Journal of Sports Science & Medicine* 18(1):13-20.

Klein, S., D.B. Allison, S.B. Heymsfield, D.E. Kelley, R.L. Leibel, C. Nonas, R. Kahn, et al. 2007. "Waist Circumference and Cardiometabolic Risk: A Consensus Statement from Shaping America's Health: Association for Weight Management and Obesity Prevention; NAASO, The Obesity Society; the American Society for Nutrition; and the American Diabetes Association." *Diabetes Care* 30(6):1647-1652. https://doi.org/10.2337/dc07-9921.

Lee, S.H. and H.S. Gong. 2020. "Measurement and Interpretation of Handgrip Strength for Research on Sarcopenia and Osteoporosis." *Journal of Bone Metabolism* 27(2):85. https://doi.org/10.11005/jbm.2020.27.2.85.

Liguori, G. and American College of Sports Medicine (ACSM). 2021. *ACSM's Guidelines for Exercise Testing and Prescription*. 11th ed. Philadelphia, PA: Wolters Kluwer.

Liu, H., W.E. Garrett, C.T. Moorman, and B. Yu. 2012. "Injury Rate, Mechanism, and Risk Factors of Hamstring Strain Injuries in Sports: A Review of the Literature." *Journal of Sport and Health Science* 1(2):92-101. https://doi.org/10.1016/j.jshs.2012.07.003.

McArdle, W.D., F.I. Katch, G.S. Pechar, L. Jacobson, and S. Ruck. 1972. "Reliability and Interrelationships between Maximal Oxygen Intake, Physical Work Capacity and Step-Test Scores in College Women." *Medicine & Science in Sports & Exercise* 4(4):182. https://doi.org/10.1249/00005768-197200440-00019.

Must, A., J. Spadano, E.H. Coakley, A.E. Field, G. Colditz, and W.H. Dietz. 1999. "The Disease Burden Associated With Overweight and Obesity." *JAMA* 282(16):1523-1529. https://doi.org/10.1001/jama.282.16.1523.

Nathan, J.A., K. Davies, and I. Swaine. 2018. "Hypermobility and Sports Injury." *BMJ Open Sport & Exercise Medicine* 4(1):e000366. https://doi.org/10.1136/bmjsem-2018-000366.

Ortega, F.B., X. Sui, C.J. Lavie, and S.N. Blair. 2016. "Body Mass Index, the Most Widely Used But Also Widely Criticized Index Would a Criterion Standard Measure of Total Body Fat Be a Better Predictor of Cardiovascular Disease Mortality?" *Mayo Clinic Proceedings* 91(4):443-455. https://doi.org/10.1016/j.mayocp.2016.01.008.

Pickering, T.G., K. Eguchi, and K. Kario. 2007. "Masked Hypertension: A Review." *Hypertension Research* 3(6):479-488. https://doi.org/10.1291/hypres.30.479.

Roberts, H.C., H.J. Denison, H.J. Martin, H.P. Patel, H. Syddall, C. Cooper, and A.A. Sayer. 2011. "A Review of the Measurement of Grip Strength in Clinical and Epidemiological Studies: Towards a Standardised Approach." *Age and Ageing* 40(4):423-429. https://doi.org/10.1093/ageing/afr051.

Sundemo, D., E.H. Senorski, L. Karlsson, A. Horvath, B. Juul-Kristensen, J. Karlsson, O.R. Ayeni, and K. Samuelsson. 2019. "Generalised Joint Hypermobility Increases ACL Injury Risk and Is Associated with Inferior Outcome after ACL Reconstruction: A Systematic Review." *BMJ Open Sport & Exercise Medicine* 5(1):e000620. https://doi.org/10.1136/bmjsem-2019-000620.

Whelton, P.K., R.M. Carey, W.S. Aronow, D.E. Casey Jr, K.J. Collins, C. Dennison Himmelfarb, et al. 2018. "2017 ACC/AHA/AAPA/ABC/ACPM/AGS/APhA/ASH/ASPC/NMA/PCNA Guideline for the Prevention, Detection, Evaluation, and Management of High Blood Pressure in Adults: A Report of the American College of Cardiology/American Heart Association Task Force on Clinical Practice Guidelines." *Hypertension* 71(6):e13-115. https://doi.org/10.1161/hyp.0000000000000065.

World Health Organization. 2000. *Obesity: Preventing and Managing the Global Epidemic*. Geneva, Switzerland: World Health Organization.

Lab 3

American College of Sports Medicine (ACSM). 2021. *ACSM's Guidelines for Exercise Testing and Prescription*. 11th ed. Baltimore: Lippincott Williams & Wilkins.

Centers for Disease Control and Prevention (CDC). 2020. "Assessing Your Weight." Last modified September 17, 2020. www.cdc.gov/healthyweight/assessing/index.html.

Centers for Disease Control and Prevention (CDC). 2021. "About Child and Teen BMI." Last modified March 17, 2021. www.cdc.gov/healthyweight/assessing/bmi/childrens_bmi/about_childrens_bmi.html.

The Cooper Institute. 2013. *Physical Fitness Assessments and Norms for Adults and Law Enforcement*. Dallas, TX: Cooper Institute.

Gibson, A., J. Holmes, R. Desautels, L. Edmonds, and L. Nuudi. 2008. "Ability of New Octapolar Bioimpedance Spectroscopy Analyzers to Predict 4-Component-Model Percentage Body Fat in Hispanic, Black, and White Adults." *American Journal of Clinical Nutrition* 87:332-338.

Gibson, A., D. Wagner, and V. Heyward. 2019. *Advanced Fitness Assessment and Exercise Prescription*. 8th ed. Champaign, IL: Human Kinetics.

Harrison, G., E. Buskirk, L. Carter, F. Johnston, T. Lohman, M. Pollock, A. Roche, and J. Wilmore. 1988. "Skinfold Thickness and Measurement Technique." In *Anthropometric Standardization Reference Manual*, edited by T. Lohman, A. Roche, and R. Martorell, 55-70. Champaign, IL: Human Kinetics.

Heyward, V. and D. Wagner. 2004. *Applied Body Composition Assessment*. 2nd ed. Champaign, IL: Human Kinetics.

Houtkooper, L., T. Lohman, S. Going, and W. Howell. 1996. "Why Bioelectrical Impedance Analysis Should be Used for Estimating Adiposity." *American Journal of Clinical Nutrition* 64(Suppl):436S-448S.

Jackson, A. and M. Pollock. 1978. "Generalized Equations for Predicting Body Density of Men." *British Journal of Nutrition* 40:497-504.

Jackson, A. and M. Pollock. 1985. "Practical Assessment of Body Composition." *The Physician and Sportsmedicine* 13:76-90.

Jackson, A., M. Pollock, and A. Ward. 1980. "Generalized Equations for Predicting Body Density of Women." *Medicine and Science in Sports and Exercise* 12:175-182.

Keys, A. and J. Brozek. 1953. "Body Fat in Adult Man." *Physiological Reviews* 88:245-325.

Keys, A. and Committee on Nutritional Anthropometry, Food and Nutrition Board, National Research Council. 1956. "Recommendations Concerning Body Composition Measurements for the Characterization of Nutritional Status." *Human Biology* 2:111-123.

Lohman, T. 1981. "Skinfolds and Body Density and Their Relation to Body Fatness: A Review." *Human Biology* 53:181-225.

Lohman, T. 1986. "Applicability of Body Composition Techniques and Constants for Children and Youths." *Exercise and Sport Sciences Review* 14:325-357.

Lohman, T.G., M.H. Slaughter, R.A. Boileau, J. Bunt, and L. Lussier. 1984. "Bone Mineral Measurements and Their Relation to Body Density in Children, Youth and Adults." *Human Biology* 56:667-679.

Ortiz, O., M. Russell, T. Daley, R. Baumgartner, M. Waki, S. Lichtman, J. Wang, R. Pierson Jr, and S. Heymsfield. 1992. "Differences in Skeletal Muscle and Bone Mineral Mass Between Black and White Females and Their Relevance to Estimates of Body Composition." *American Journal of Clinical Nutrition* 55:8-13.

Pascale, L., M. Grossman, H. Sloane, and T. Frankel. 1956. "Correlations Between Thickness of Skinfolds and Body Density in 88 Soldiers." *Human Biology* 2:165-176.

Pollock, M. and A. Jackson. 1984. "Research Progress in Validation of Clinical Methods of Assessing Body Composition." *Medicine & Science in Sports & Exercise* 16:606-613.

Siri, W.E. 1956. "The Gross Composition of the Body." In *Advances in Biological and Medical Physics*, edited by C.A. Tobias and J.H. Lawrence, 239-280. New York: Academic Press.

Wagner, D. and V. Heyward. 2001. "Validity of Two-Component Models for Estimating Body Fat of Black Men." *Journal of Applied Physiology* 90:649-656.

Lab 4

Brožek, J., F. Grande, J.T. Anderson, and A. Keys. 1963. "Densiometric Analysis of Body Composition: Revision of Some Quantitative Assumptions." *Annals of the New York Academy of Sciences* 11(1):113-140. https://doi.org/10.1111/j.1749-6632.1963.tb17079.x.

Burge, R., B. Dawson-Hughes, D.H. Solomon, J.B. Wong, A. King, and A. Tosteson. 2007. "Incidence and Economic Burden of Osteoporosis-Related Fractures in the United States, 2005-2025." *Journal of Bone and Mineral Research* 22(3):465-475. https://doi.org/10.1359/jbmr.061113.

Fields, D.A., M.I. Goran, and M.A. McCrory. 2002. "Body-Composition Assessment via Air-Displacement Plethysmography in Adults and Children: A Review." *The American Journal of Clinical Nutrition* 75(3):453-467. https://doi.org/10.1093/ajcn/75.3.453.

Kohrt, W.M. 1997. "Dual-Energy X-Ray Absorptiometry: Research Issues and Equipment." In *Emerging Technologies for Nutrition Research: Potential for Assessing Military Performance Capability*, by Institute of Medicine (US)

Committee on Military Nutrition Research, edited by S.J. Carlson-Newberry and R.B. Costello. Washington, DC: National Academies Press (US).

Moon, J.R., J.R. Stout, A.A. Walter, A.E. Smith, M.S. Stock, T.J. Herda, V.D. Sherk, et al. 2011. "Mechanical Scale and Load Cell Underwater Weighing: A Comparison of Simultaneous Measurements and the Reliability of Methods." *Journal of Strength and Conditioning Research* 25(3):652-661. https://doi.org/10.1519/jsc.0b013e3181e99c2d.

Morrow Jr, J.R., A.S. Jackson, P.W. Bradley, and G.H. Hartung. 1986. "Accuracy of Measured and Predicted Lung Volume on Body Density Measurement." *Medicine and Science in Sports and Exercise* 18(6):647-652.

Siri, W.E. 1961. "Body Composition from Fluid Spaces and Density: Analysis of Methods." In *Techniques for Measuring Body Composition*, edited by J. Brozek and A. Hanschel, 223-244. Washington, DC: National Academy of Science.

Varacallo, M.A., E.J. Fox, E.M. Paul, S.E. Hassenbein, and P.M. Warlow. 2013. "Patients' Response Toward an Automated Orthopedic Osteoporosis Intervention Program." *Geriatric Orthopaedic Surgery & Rehabilitation* 4(3):89-98. https://doi.org/10.1177/2151458513502039.

Lab 5

Aragon, A.A., B.J. Schoenfeld, R. Wildman, S. Kleiner, T. VanDusseldorp, L. Taylor, C.P. Earnest, et al. 2017. "International Society of Sports Nutrition Position Stand: Diets and Body Composition." *Journal of the International Society of Sports Nutrition* 14(1):16.

Bianchini, F., R. Kaaks, and H. Vainio. 2002. "Overweight, Obesity, and Cancer Risk." *The Lancet Oncology* 3(9):565-574.

Biener, A., J. Cawley, and C. Meyerhoefer. 2018. "The Impact of Obesity on Medical Care Costs and Labor Market Outcomes in the US." *Clinical Chemistry* 64(1):108-117.

Centers for Disease Control and Prevention (CDC). 2004. "Prevalence of Overweight and Obesity among Adults with Diagnosed Diabetes—United States, 1988-1994 and 1999-2002." *MMWR. Morbidity and Mortality Weekly Report* 53(45):1066-1068.

Donnelly, J.E., S.N. Blair, J.M. Jakicic, M.M. Manore, J.W. Rankin, B.K. Smith, and American College of Sports Medicine. 2009. "American College of Sports Medicine Position Stand. Appropriate Physical Activity Intervention Strategies for Weight Loss and Prevention of Weight Regain for Adults." *Medicine and Science in Sports and Exercise* 41(2):459-471.

Harris, J.A. and F.G. Benedict. 1918. "A Biometric Study of Human Basal Metabolism." *Proceedings of the National Academy of Sciences* 4(12):370-373.

Harsha, D.W. and G.A. Bray. 2008. "Weight Loss and Blood Pressure Control (Pro)." *Hypertension* 51(6):1420-1425.

Jackson, A.S. and M.L. Pollock. 1978. "Generalized Equations for Predicting Body Density of Men." *British Journal of Nutrition* 40(3):497-504.

Jahangir, E., A. De Schutter, and C.J. Lavie. 2014. "The Relationship Between Obesity and Coronary Artery Disease." *Translational Research* 164(4):336-344.

Khaodhiar, L., K.C. McCowen, and G.L. Blackburn. 1999. "Obesity and Its Comorbid Conditions." *Clinical Cornerstone* 2(3):17-31.

Levine, S., E. Malone, A. Lekiachvili, and P. Briss. 2019. "Health Care Industry Insights: Why the Use of Preventive Services Is Still Low." *Preventing Chronic Disease* 16:E30.

Macdiarmid, J. and J. Blundell. 1998. "Assessing Dietary Intake: Who, What and Why of Under-reporting." *Nutrition Research Reviews* 11(2):231-253.

Mifflin, M.D., S.T. St Jeor, L.A. Hill, B.J. Scott, S.A. Daugherty, and Y.O. Koh. 1990. "A New Predictive Equation for Resting Energy Expenditure in Healthy Individuals." *The American Journal of Clinical Nutrition* 51(2):241-247.

Ruggiero, C. and L. Ferrucci. 2006. "The Endeavor of High Maintenance Homeostasis: Resting Metabolic Rate and the Legacy of Longevity." *The Journals of Gerontology: Series A* 61(5):466-473.

Tsai, A.G., D.F. Williamson, and H.A. Glick. 2011. "Direct Medical Cost of Overweight and Obesity in the USA: A Quantitative Systematic Review." *Obesity Reviews* 12(1):50-61.

World Health Organization (WHO). 2000. "Obesity: Preventing and Managing the Global Epidemic. Report of a WHO Consultation." *World Health Organization Technical Report Series* 894:i-xii, 1-253.

World Health Organization (WHO). 2021. "Obesity and Overweight." Last modified June 9, 2021. www.who.int/news-room/fact-sheets/detail/obesity-and-overweight.

Lab 6

American College of Sports Medicine (ACSM). 2021. *ACSM's Guidelines for Exercise Testing and Prescription*. 11th ed. Baltimore: Lippincott Williams & Wilkins.

Astrand, P., K. Rodahl, H. Dahl, and S. Stromme. 2003. *Textbook of Work Physiology*. 4th ed. Champaign, IL: Human Kinetics.

Beltz, N., A. Gibson, J. Janot, L. Kravitz, C. Mermier, and L. Dalleck. 2016. "Graded Exercise Testing Protocols for the Determination of VO_2max: Historical Perspectives, Progress and Future Considerations." *Journal of Sports Medicine* 2016:1-12.

Gibson, A., D. Wagner, and V. Heyward. 2019. *Advanced Fitness Assessment and Exercise Prescription*. 8th ed. Champaign, IL: Human Kinetics.

Moore, C.C. 2019. "Development and Cross-Validation of a Cadence-Based Metabolic Equation for Walking." Master's thesis, University of Massachusetts Amherst.

Wagner, J., M. Niemeyer, D. Infanger, T. Hinrichs, L. Streese, H. Hanssen, J. Myers, et al. 2020. "New Data-based Cut-offs for Maximal Exercise Criteria across the Lifespan." *Medicine and Science in Sports and Exercise* 52:1915-1923.

Lab 7

Ainsworth, B., W. Haskell, A. Leon, D. Jacobs, H. Montoye, J. Sallis, and R. Paffenbarger. 1993. "Compendium of Physical Activities: Classification of Energy Costs of Human Physical Activities." *Medicine and Science in Sports and Exercise* 25:71-80.

American College of Sports Medicine (ACSM). 2021. *ACSM's Guidelines for Exercise Testing and Prescription.* 11th ed. Baltimore: Lippincott Williams & Wilkins.

Astrand, P., K. Rodahl, H. Dahl, and S. Stromme. 2003. *Textbook of Work Physiology.* 4th ed. Champaign, IL: Human Kinetics.

Barry, V., M. Baruth, M. Beets, J. Durstine, J. Liu, and S. Blair. 2014. "Fitness vs. Fatness on All-Cause Mortality: A Meta-Analysis." *Progress in Cardiovascular Diseases* 56:382-390.

Beekley, M., W. Brechue, D. DeHoyos, L. Garzarella, G. Werber-Zion, and M. Pollock. 2004. "Cross-validation of the YMCA Submaximal Cycle Ergometer Test to Predict $\dot{V}O_2$max." *Research Quarterly for Exercise and Sport* 75:337-342.

Blair, S., H. Kohl, R. Paffenbarger, D. Clark, K. Cooper, and L. Gibbons. 1989. "Physical Fitness and All-Cause Mortality. A Prospective Study of Healthy Men and Women." *Journal of the American Medical Association* 262:2395-2401.

Borg, G. 1982. "Psychophysical Bases of Perceived Exertion." *Medicine and Science in Sports and Exercise* 14:377-381.

Bull, F., S. Al-Ansari, S. Biddle, K. Borodulin, M. Buman, G. Cardon, C. Carty, et al. 2020. "World Health Organization 2020 Guidelines on Physical Activity and Sedentary Behavior." *British Journal of Sports Medicine* 54:1451-1462.

Bunc, V., P. Hofmann, H. Leitner, and G. Gaisi. 1995. "Verification of the Heart Rate Threshold." *European Journal of Applied Physiology* 70:263-269.

Centers for Disease Control and Prevention (CDC). 2021. "Lack of Physical Activity." Last modified September 25, 2019. www.cdc.gov/chronicdisease/resources/publications/factsheets/physical-activity.htm.

Conconi, F., M. Ferrari, P. Ziglio, P. Droghetti, and L. Codeca. 1982. "Determination of the Anaerobic Threshold by a Noninvasive Field Test in Runners." *Journal of Applied Physiology* 52:869-873.

Dehart-Beverly, M., C. Foster, J. Porcari, D. Fater, and R. Mikat. 2000. "Relationship Between the Talk Test and Ventilatory Threshold." *Clinical Exercise Physiology* 2:34-38.

Ebbeling, C., A. Ward, E. Puleo, J. Widrick, and J. Rippe. 1991. "Development of a Single-Stage Submaximal Treadmill Walking Test." *Medicine and Science in Sports and Exercise* 23:966-973.

Fox, S., J. Naughton, and W. Haskell. 1971. "Physical Activity and the Prevention of Coronary Heart Disease." *Annals of Clinical Research* 3:404-432.

Garber, C., B. Blissmer, M. Deschenes, B. Franklin, M. Lamonte, I. Lee, D. Nieman, and D. Swain. 2011. "American College of Sports Medicine position stand. Quantity and Quality of Exercise for Developing and Maintaining Cardiorespiratory, Musculoskeletal, and Neuromotor Fitness in Apparently Healthy Adults: Guidance for Prescribing Exercise." *Medicine and Science in Sports and Exercise* 43:1334-1359.

Gellish, R., B. Goslin, R. Olson, A. McDonald, G. Russi, and V. Moudgil. 2007. "Longitudinal Modeling of the Relationship Between Age and Maximal Heart Rate." *Medicine and Science in Sports and Exercise* 39:822-829.

George, J., P. Vehrs, P. Allsen, G. Fellingham, and A. Fisher. 1993a. "$\dot{V}O_2$max Estimation from a Submaximal 1-Mile Track Jog for Fit College-Age Individuals." *Medicine and Science in Sports and Exercise* 25:401-406.

George, J., P. Vehrs, P. Allsen, G. Fellingham, and A. Fisher. 1993b. "Development of a Submaximal Treadmill Jogging Test for Fit College-Aged Individuals." *Medicine and Science in Sports and Exercise* 25:643-647.

George, J., P. Vehrs, G. Babcock, M. Etchie, T. Chinevere, and G. Fellingham. 2000. "A Modified Submaximal Cycle Ergometer Test Designed to Predict Treadmill $\dot{V}O_2$max." *Measurement in Physical Education and Exercise Science* 4:229-243.

Gibson, A., D. Wagner, and V. Heyward. 2019. *Advanced Fitness Assessment and Exercise Prescription.* 8th ed. Champaign, IL: Human Kinetics.

Golding, L. 2000. *The Y's Way to Physical Fitness.* Champaign, IL: Human Kinetics.

Hofmann, P., V. Bunc, H. Leitner, R. Pokan, and G. Gaisl. 1994. "Heart Rate Threshold Related to Lactate Turn Point and Steady-State Exercise on a Cycle Ergometer." *European Journal of Applied Physiology* 69:132-139.

Hofmann, P., R. Pokan, F.-J. Seibert, R. Zweiker, and P. Schmid. 1997. "The Heart Rate Performance Curve During Incremental Cycle Ergometer Exercise in Healthy Young Male Subjects." *Medicine and Science in Sports and Exercise* 29:762-768.

Kaminsky, L., R. Arena, Ø. Ellingsen, M. Harber, J. Myers, C. Ozemek, and R. Ross. 2019. "Cardiorespiratory Fitness and Cardiovascular Disease: The Past, Present, and Future." *Progress in Cardiovascular Diseases* 62:86-93.

Kaminsky, L., R. Arena, and J. Myers. 2015. "Reference Standards for Cardiorespiratory Fitness Measured with Cardiopulmonary Exercise Testing: Data from the Fitness Registry and the Importance of Exercise National Database." *Mayo Clinic Proceedings* 90:1515-1523.

Kaminsky, L., M. Imboden, R. Arena, and J. Myers. 2017. "Reference Standards for Cardiorespiratory Fitness Measured with Cardiopulmonary Exercise Testing Using Cycle Ergometry: Data from the Fitness Registry and the Importance of Exercise National Database (FRIEND) Registry." *Mayo Clinic Proceedings* 90:1515-1523.

Kline, G., J. Porcari, R. Hintermeister, P. Freedson, A. Ward, R. McCarron, J. Ross, and J. Rippe. 1987. "Estimation of $\dot{V}O_2$max From a One-mile Track Walk, Gender, Age, and Body Weight." *Medicine and Science in Sports and Exercise* 19:253-259.

Kohrt, W., R. Spina, J. Hollozy, and A. Ehsani. 1998. "Prescribing Exercise Intensity for Older Women." *Journal of the American Geriatrics Society* 46:129-133.

Kraus, W., C. Torgan, B. Duscha, J. Norris, S. Brown, F. Cobb, C. Bales, et al. 2001. "Studies of a Targeted Risk Reduction Intervention Through Defined Exercise (STRRIDE)." *Medicine and Science in Sports and Exercise* 33:1774-1784.

Larsen, G., J. George, J. Alexander, G. Fellingham, S. Aldana, and A. Parcell. 2002. "Prediction of Maximum Oxygen Consumption from Walking, Jogging, or Running." *Research Quarterly for Exercise and Sport* 73:66-72.

Lee, D., E. Artero, X. Sui, and S. Blair. 2010. "Mortality Trends in the General Population: The Importance of Cardiorespiratory Fitness." *Journal of Psychopharmacology* 24:27-35.

Myers, J., P. McAuley, C. Lavie, J. Despres, R. Arena, and P. Kokkinos. 2015. "Physical Activity and Cardiorespiratory Fitness as Major Markers of Cardiovascular Risk: Their Independent and Interwoven Importance to Health Status." *Progress in Cardiovascular Diseases* 57:306-314.

Porcari, J., K. Falck-Wiese, S. Suckow-Stenger, J. Turek, A. Wargowsky, M. Cress, S. Doberstein, et al. 2018. "Comparison of the Talk Test and Percent Heart Rate Reserve for Exercise Prescription." *Kinesiology* 50:3-10.

Robergs, R. and R. Landwehr. 2002. "The Surprising History of the 'HRmax = 220-age' Equation." *Journal of Exercise Physiology Online* 5:1-10.

Robergs, R. and S. Roberts. 1997. *Exercise Physiology: Exercise, Performance and Clinical Applications*. St. Louis, MO: Mosby.

Ross, R., S. Blair, R. Arena, T. Church, J. Després, B. Franklin, W. Haskell, et al. 2016. "Importance of Assessing Cardiorespiratory Fitness in Clinical Practice: A Case for Fitness as a Clinical Vital Sign: A Scientific Statement from the American Heart Association." *Circulation* 134:e653-699.

Sarzynski, M., T. Rankinen, C. Earnest, A. Leon, D. Rao, J. Skinner, and C. Bouchard. 2013. "Measured Maximal Heart Rates Compared to Commonly Used Age-Based Prediction Equations in the Heritage Family Study." *American Journal of Human Biology* 25:695-701.

Taylor, W., J. George, P. Allsen, P. Vehrs, R. Hager, and M. Roberts. 2002. "Estimation of $\dot{V}O_2$max from a 1.5 mile Endurance Test." *Medicine and Science in Sports and Exercise* 35:S257 [abstract].

Tudor-Locke, C., C. Craig, W. Brown, S. Clemes, K. De Cocker, B. Giles-Corti, Y. Hatano, et al. 2011. "How Many Steps/Day are Enough? For Adults." *International Journal of Behavioral Nutrition and Physical Activity* 8:79.

Tudor-Locke, C., H. Han, E. Aguiar, T. Barreira, J. Schuna, M. Kang, and D. Rowe. 2018. "How Fast is Fast Enough? Walking Cadence (Steps/Min) as a Practical Estimate of Intensity in Adults: A Narrative Review." *British Journal of Sports Medicine* 52:776-788.

U.S. Department of Health and Human Services (USDHHS). 2018. *Physical Activity Guidelines for Americans*. 2nd ed. Washington, DC: U.S. Department of Health and Human Services.

Vella, C. and R. Robergs. 2005. "Non-Linear Relationships Between Central Cardiovascular Variables and $\dot{V}O_2$ During Incremental Cycling Exercise in Endurance-Trained Individuals." *Journal of Sports Medicine and Physical Fitness* 45:452-459.

Weatherwax, R., N. Harris, A. Kilding, and L. Dalleck. 2019. "Incidence of $\dot{V}O_2$max Responders to Personalized Versus Standardized Exercise Prescription." *Medicine and Science in Sports and Exercise* 51:681-691.

Zeni, A., M. Hoffman, and P. Clifford. 1996. "Relationships Among Heart Rate, Lactate Concentration, and Perceived Effort for Different Types of Rhythmic Exercise in Women." *Archives of Physical Medicine and Rehabilitation* 77:237-241.

Lab 8

Bassett, D.R. Jr and E.T. Howley. 2000. "Limiting Factors For Maximal Oxygen Uptake and Determinants of Endurance Performance." *Medicine and Science in Sports and Exercise* 32:70-84.

Beaver, W.L., K. Wasserman, and B.J. Whipp. 1986. "A New Method for Detecting Anaerobic Threshold by Gas Exchange." *Journal of Applied Physiology* 60:2020-2027.

Beltz, N.M., A.L. Gibson, J.M. Janot, L. Kravitz, C.M. Mermier, and L.C. Dalleck. 2016. "Graded Exercise Testing Protocols for the Determination of VO_2max: Historical Perspectives, Progress, and Future Considerations." *Journal of Sports Medicine* 2016:3968393. https://doi.org/10.1155/2016/3968393.

Berling, J., C. Foster, M. Gibson, S. Doberstein, and J. Porcari. 2006. "The Effect of Handrail Support on Oxygen Uptake During Steady-State Treadmill Exercise." *Journal of Cardiopulmonary Rehabilitation* 26:391-394.

Borg, G. 1982. "Psychophysical Bases of Perceived Exertion." *Medicine and Science in Sports and Exercise* 14:377-381.

Bruce, R.A., J.R. Blackmon, J.W. Jones, and G. Strait. 1963. "Exercise Testing in Adult Normal Subjects and Cardiac Patients." *Pediatrics* 32:742-756.

Buchfuhrer, M.J., J.E. Hanson, T.E. Robinson, D.Y. Sue, K. Wasserman, and B.J. Whipp. 1983. "Optimizing the Exercise Protocol for Cardiopulmonary Assessment." *Journal of Applied Physiology: Respiratory, Environmental and Exercise Physiology* 55:1558-1564.

Eston, R.G., K.L. Lamb, G. Parfitt, and N. King. 2005. "The Validity of Predicting Maximal Oxygen Uptake From a Perceptually-Regulated Graded Exercise Test." *European Journal of Applied Physiology* 94:221-227.

Gellish, R., B. Goslin, R. Olson, A. McDonald, G. Russi, and V. Moudgil. 2007. "Longitudinal Modeling of the Relationship Between Age and Maximal Heart Rate." *Medicine and Science in Sports and Exercise* 39:822-829.

Mauger, A.R. and N. Sculthorpe. 2012. "A New VO_2max Protocol Allowing Self-Pacing in Maximal Incremental Exercise." *British Journal of Sports Medicine* 46:59-63.

Myers, J., N. Buchanan, D. Walsh, M. Kraemer, P. McAuley, M. Hamilton-Wessler, and V.F. Froelich. 1991. "Comparison of the Ramp Versus Standard Exercise Protocols." *Journal of the American College of Cardiology* 17:1334-1342.

Vainshelboim, B., R. Arena, L.A. Kaminsky, and J. Myers. 2020. "Reference Standards for Ventilatory Threshold Measures with Cardiopulmonary Exercise Testing. The Fitness Registry and the Importance of Exercise: A National Database." *Chest* 157:1531-1537.

Weatherwax, R.M., N.K. Harris, A.E. Kilding, and L.C. Dalleck. 2019. "Incidence of VO$_2$max Responders versus Standardized Exercise Prescription." *Medicine and Science in Sports and Exercise* 51:681-691.

Lab 9

American College of Sports Medicine (ACSM). 2021. *ACSM's Guidelines for Exercise Testing and Prescription.* 11th ed. Baltimore: Lippincott Williams & Wilkins.

Beam, W. and G. Adams. 2011. *Exercise Physiology Laboratory Manual.* 6th ed. New York: McGraw Hill.

Beardsley, C. and J. Škarabot. 2015. "Effects of Self-Myofascial Release: A Systematic Review." *Journal of Bodywork and Movement Therapies* 19:747-758.

Behm, D., A. Blazevich, A. Kay, and M. McHugh. 2016. "Acute Effects of Muscle Stretching on Physical Performance, Range of Motion, and Injury Incidence in Healthy Active Individuals: A Systematic Review." *Applied Physiology, Nutrition, and Metabolism* 41:1-11.

Behm, D. and A. Chaouachi. 2011. "A Review of the Acute Effects of Static and Dynamic Stretching on Performance." *European Journal of Applied Physiology* 111:2633-2651.

Borges, M., D. Medeiros, B. Minotto, and C. Lima. 2018. "Comparison Between Static Stretching and Proprioceptive Neuromuscular Facilitation on Hamstring Flexibility: Systematic Review and Meta-Analysis." *European Journal of Physiotherapy* 20:12-19.

Cheatham, S., M. Kolber, M. Cain, and M. Lee. 2015. "The Effects of Self-Myofascial Release Using a Foam Roll or Roller Massager on Joint Range of Motion, Muscle Recovery, and Performance: A Systematic Review." *International Journal of Sports Physical Therapy* 10:827-838.

Garber, C., B. Blissmer, M. Deschenes, B. Franklin, M. Lamonte, I. Lee, D. Nieman, and D. Swain. 2011. "American College of Sports Medicine Position Stand. Quantity and Quality of Exercise for Developing and Maintaining Cardiorespiratory, Musculoskeletal, and Neuromotor Fitness in Apparently Healthy Adults: Guidance for Prescribing Exercise." *Medicine and Science in Sports and Exercise* 43:1334-1359.

Gibson, A., D. Wagner, and V. Heyward. 2019. *Advanced Fitness Assessment and Exercise Prescription.* 8th ed. Champaign, IL: Human Kinetics.

Glei, D., N. Goldman, C. Ryff, and M. Weinstein. 2019. "Physical Function in U.S. Older Adults Compared with Other Populations: A Multinational Study." *Journal of Aging and Health* 31:1067-1084.

Gordon, N., M. Gulanick, F. Costa, G. Fletcher, B. Franklin, E. Roth, and T. Shephard; American Heart Association Council on Clinical Cardiology, Subcommittee on Exercise, Cardiac Rehabilitation, and Prevention; the Council on Cardiovascular Nursing; the Council on Nutrition, Physical Activity, and Metabolism; and the Stroke Council. 2004. "Physical Activity and Exercise Recommendations for Stroke Survivors: An American Heart Association Scientific Statement from the Council on Clinical Cardiology, Subcommittee on Exercise, Cardiac Rehabilitation, and Prevention; the Council on Cardiovascular Nursing; the Council on Nutrition, Physical Activity, and Metabolism; and the Stroke Council." *Circulation* 109:2031-2041.

Greene, W. and J. Heckman, eds. 1994. *The Clinical Measurement of Joint Motion.* 1st ed. Rosemont, IL: American Academy of Orthopedic Surgeons.

Grenier, S., C. Russell, and S. McGill. 2003. "Relationships Between Lumbar Flexibility, Sit-and-Reach Test, and a Previous History of Low Back Discomfort in Industrial Workers." *Canadian Journal of Applied Physiology* 28:165-177.

Hoeger, W. and S. Hoeger. 2015. *Lifetime Physical Fitness & Wellness: A Personalized Program.* 13th ed. Boston, MA: Cengage Learning.

Hoeger, W. and D. Hopkins. 1992. "A Comparison of the Sit and Reach and the Modified Sit and Reach in the Measurement of Flexibility in Women." *Research Quarterly for Exercise and Sport* 63:191-195.

Jackson, A., J. Morrow, P. Brill, H. Kohl, N. Gordon, and S. Blair. 1998. "Relations of Sit-Up and Sit-and-Reach Tests to Low Back Pain in Adults." *Journal of Orthopaedic & Sports Physical Therapy* 27:22-26.

Jones, B. and J. Knapik. 1999. "Physical Training and Exercise-related Injuries. Surveillance, Research and Injury Prevention in Military Populations." *Sports Medicine* 27:111-125.

Kay, A. and A. Blazevich. 2012. "Effect of Acute Static Stretch on Maximal Muscle Performance: A Systematic Review." *Medicine and Science in Sports and Exercise* 44:154-164.

Kendall, F., E. McCreary, P. Provance, M. Rodgers, and W. Romani. 2005. *Muscles: Testing and Function with Posture and Pain.* 5th ed. Philadelphia, PA: Lippincott Williams & Wilkins.

Kokkonen, J., A. Nelson, C. Eldredge, and B. Winchester. 2007. "Chronic Static Stretching Improves Exercise Performance." *Medicine and Science in Sports and Exercise* 39:1825-1831.

Kolber, M., M. Pizzini, A. Robinson, D. Yanez, and W. Hanney. 2013. "The Reliability and Concurrent Validity of Measurements Used to Quantify Lumbar Spine Mobility: An Analysis of an iPhone® Application and Gravity Based Inclinometry." *International Journal of Sports Physical Therapy* 8:129-137.

Konrad, A., S. Stafilidis, and M. Tilp. 2017. "Effects of Acute Static, Ballistic, and PNF Stretching Exercise on The Muscle and Tendon Tissue Properties." *Scandinavian Journal of Medicine & Science in Sports* 27:1070-1080.

Lemmink, K., H. Kemper, M. de Greef, P. Rispens, and M. Stevens. 2003. "The Validity of the Sit-and-Reach Test and the Modified Sit-and-Reach Test in Middle-Aged to Older Men and Women." *Research Quarterly for Exercise and Sport* 74:331-336.

Lempke, L., R. Wilkinson, C. Murray, and J. Stanek. 2018. "The Effectiveness of PNF Versus Static Stretching on Increasing Hip-Flexion Range of Motion." *Journal of Sports Rehabilitation* 27:289-294.

López-Miñarro, P., P. Andújar, and P. Rodríguez-García. 2009. "A Comparison of the Sit-and-Reach Test and the Back-saver Sit-and-Reach Test in University Students." *Journal of Sports Science and Medicine* 8:116-122.

Martin, S., A. Jackson, J. Morrow, and W. Liemohn. 1998. "The Rationale for the Sit and Reach Test Revisited." *Measurement in Physical Education and Exercise Science* 2:85-92.

Mayer, T., A. Tencer, S. Kristoferson, and V. Mooney. 1984. "Use of Noninvasive Techniques for Quantification of Spinal Range-of-Motion in Normal Subjects and Chronic Low-back Dysfunction Patients." *Spine* 9:588-595.

Merrill, S. 2015. "Flexibility Training." In *Exercise Physiology*, edited by J. Porcari, C. Bryant, and F. Comana, 447-466. Philadelphia, PA: F.A. Davis.

Ng, J., V. Kippers, C. Richardson, and M. Parnianpour. 2001. "Range of Motion and Lordosis of the Lumbar Spine: Reliability of Measurement and Normative Values." *Spine* 26:53-60.

Samo, D., S. Chen, A. Crampton, E. Chen, K. Conrad, L. Egan, and J. Mitton. 1997. "Validity of Three Lumbar Sagittal Motion Measurement Methods: Surface Inclinometers Compared with Radiographs." *Journal of Occupational and Environmental Medicine* 39:209-216.

Saur, P., F. Ensink, K. Frese, D. Seeger, and J. Hildebrandt. 1996. "Lumbar Range of Motion: Reliability and Validity of the Inclinometer Technique in the Clinical Measurement of Trunk Flexibility." *Spine* 21:1332-1338.

Schleip, R. and D. Müller. 2013. "Training Principles for Fascial Connective Tissues: Scientific Foundation and Suggested Practical Applications." *Journal of Bodywork and Movement Therapies* 17:103-115.

Sharman, M., A. Cresswell, and S. Riek. 2006. "Proprioceptive Neuromuscular Facilitation Stretching: Mechanisms and Clinical Implications." *Sports Medicine* 36:929-939.

Shrier, I. 2004. "Does Stretching Improve Performance? A Systematic and Critical Review of the Literature." *Clinical Journal of Sports Medicine* 14:267-273.

Shultz, S., P. Houglum, and D. Perrin. 2015. *Examination of Musculoskeletal Injuries.* 4th ed. Champaign, IL: Human Kinetics.

Wilke, J., A. Müller, F. Giesche, G. Power, H. Ahmedi, and D. Behm. 2020. "Acute Effects of Foam Rolling on Range of Motion in Healthy Adults: A Systematic Review with Multilevel Meta-analysis." *Sports Medicine* 50:387-402.

Lab 10

American College of Sports Medicine (ACSM). 2021. *ACSM's Guidelines for Exercise Testing and Prescription.* 11th ed. Baltimore: Lippincott Williams & Wilkins.

Behm, D. and K. Anderson. 2006. "The Role of Instability with Resistance Training." *Journal of Strength and Conditioning Research* 20:716-722.

Behm, D. and J. Colado. 2012. "The Effectiveness of Resistance Training Using Unstable Surfaces and Devices for Rehabilitation." *International Journal of Sports Physical Therapy* 7:226-241.

Behm, D., T. Muehlbauer, A. Kibele, and U. Granacher. 2015. "Effects of Strength Training Using Unstable Surfaces on Strength, Power and Balance Performance Across the Lifespan: A Systematic Review and Meta-analysis." *Sports Medicine* 45:1645-1669.

Behm, D. and J. Sanchez. 2013. "Instability Resistance Training Across the Exercise Continuum." *Sports Health* 5:500-503.

Beltz, N., T. Nuñez, and J. Janot. 2019. "Effect of Functional Resistance Training on Movement Outcomes in Young Adults." *Journal of Exercise Physiology Online* 22:227-238.

Garber, C., B. Blissmer, M. Deschenes, B. Franklin, M. Lamonte, I. Lee, D. Nieman, and D. Swain. 2011. "American College of Sports Medicine Position Stand. Quantity and Quality of Exercise for Developing and Maintaining Cardiorespiratory, Musculoskeletal, and Neuromotor Fitness in Apparently Healthy Adults: Guidance for Prescribing Exercise." *Medicine and Science in Sports and Exercise* 43:1334-1359.

Gibson, A., D. Wagner, and V. Heyward. 2019. *Advanced Fitness Assessment and Exercise Prescription.* 8th ed. Champaign, IL: Human Kinetics.

Golding, L. 2000. *The Y's Way to Physical Fitness.* Champaign, IL: Human Kinetics.

Hayden, J., M. van Tulder, and G. Tomlinson. 2005. "Systematic Review: Strategies for Using Exercise Therapy to Improve Outcomes in Chronic Low Back Pain." *Annals of Internal Medicine* 142:776-785.

Janot, J., T. Heltne, C. Welles, J. Riedel, H. Anderson, A. Howard, and S. Myhre. 2013. "Effects of TRX Versus Traditional Resistance Training on Measures of Muscular Performance in Young and Middle-Aged Adults." *Journal of Fitness Research* 2:23-38.

Kell, R. and G. Asmundson. 2009. "A Comparison of Two Forms of Periodized Exercise Rehabilitation Programs in the Management of Chronic Nonspecific Low-Back Pain." *Journal of Strength and Conditioning Research* 23:513-523.

Magyari, P., ed. 2018. *ACSM's Resources for the Exercise Physiologist. A Practical Guide for the Health Fitness Professional.* 2nd ed. Philadelphia, PA: Wolters Kluwer.

Mayer, J., W. Quillen, J. Verna, R. Chen, P. Lunseth, and S. Dagenais. 2015. "Impact of a Supervised Worksite Exercise

Program on Back and Core Muscular Endurance in Fire-fighters." *American Journal of Health Promotion* 29:165-172.

McGill, S. 2015. *Low Back Disorders: Evidence-Based Prevention and Rehabilitation*. 3rd ed. Champaign, IL: Human Kinetics.

McGill, S., A. Childs, and C. Liebenson. 1999. "Endurance Times for Low Back Stabilization Exercises: Clinical Targets for Testing and Training from a Normal Database." *Archives of Physical Medicine and Rehabilitation* 80:941-944.

McGill, S., S. Grenier, M. Bluhm, R. Preuss, S. Brown, and C. Russell. 2003. "Previous History of LBP with Work Loss is Related to Lingering Deficits in Biomechanical, Physiological, Personal, Psychosocial and Motor Control Characteristics." *Ergonomics* 46:731-746.

Moon, H., K. Choi, D. Kim, H. Kim, Y. Cho, J. Kim, and Y. Choi. 2013. "Effect of Lumbar Stabilization and Dynamic Lumbar Strengthening Exercises in Patients with Chronic Low Back Pain." *Annals of Rehabilitation Medicine* 37:110-117.

Nuñez, T., F. Amorim, J. Janot, C. Mermier, R. Rozenek, and L. Kravitz. 2017. "Circuit Weight Training: Acute and Chronic Effects on Healthy and Clinical Populations." *Journal of Sport and Human Performance* 5:1-21.

Rainville, J., C. Hartigan, E. Martinez, J. Limke, C. Jouve, and M. Finno. 2004. "Exercise as a Treatment for Chronic Low Back Pain." *The Spine Journal* 4:106-115.

Ratamess, N., B. Alvar, T. Evetovich, T. Housh, W. Kibler, W. Kraemer, and N. Triplett. 2009. "American College of Sports Medicine Position Stand. Progression Models in Resistance Training for Healthy Adults." *Medicine and Science in Sports and Exercise* 41:687-708.

YMCA of the USA. 2000. *YMCA Fitness Testing and Assessment Manual*. 4th ed. Champaign, IL: Human Kinetics.

Lab 11

American College of Sports Medicine (ACSM). 2021a. *ACSM's Guidelines for Exercise Testing and Prescription*. 11th ed. Baltimore: Lippincott Williams & Wilkins.

American College of Sports Medicine (ACSM). 2021b. *ACSM's Resources for the Exercise Physiologist*. 2nd ed. Baltimore: Lippincott Williams & Wilkins.

Baechle, T.R. and R.W. Earle. 2020. *Weight Training: Steps to Success*. 5th ed. Champaign, IL: Human Kinetics.

Brzycki, M. 1993. "Strength Testing: Predicting a One-Repo Max from Reps-to-Fatigue." *Journal of Health, Physical Education, Recreation, and Dance* 64:88-90.

Gibson, A., D. Wagner, and V. Heyward. 2019. *Advanced Fitness Assessment and Exercise Prescription*. 8th ed. Champaign, IL: Human Kinetics.

Haff, G.G. and N.T. Triplett. 2016. *Essentials of Strength Training and Conditioning*. 4th ed. Champaign, IL: Human Kinetics.

Helms, E.R., J. Cronin, A. Storey, and M.C. Zourdos. 2016. "Application of the Repetitions in Reserve-Based Rating

of Perceived Exertion Scale for Resistance Training." *Journal of Strength and Conditioning Research* 38:42-49.

LeSuer, D.A., J.H. McCormick, J.L. Mayhew, R.L. Wasserstein, and M.D. Arnold. 1997. "The Accuracy of Prediction Equations for Estimating 1-RM Performance in the Bench Press, Squat, and Deadlift." *Journal of Strength and Conditioning Research* 11:211-213.

Mayhew, J.L., J.R. Ware, and J.L. Prinster. 1993. "Using Lift Repetitions to Predict Muscular Strength in Adolescent Males." *Strength and Conditioning Journal* 15:35-38.

Pacifico, J., M.A.J. Geerlings, E.M. Reijnierse, C. Phassouliotis, W.K. Lim, and A.B. Maier. 2020. "Prevalence of Sarcopenia as a Comorbid Disease: A Systematic Review and Meta-Analysis." *Experimental Gerontology* 131:110801.

Peterson, M.D., M.R. Rhea, and B.A. Alvar. 2005. "Applications of the Dose-Response for Muscular Strength Development: A Review of Meta-Analytic Efficacy and Reliability for Designing Training Prescription." *Journal of Strength and Conditioning Research* 19:950-958.

Pizzigalli, P., A. Filippini, A. Alberto, S. Ahmaidi, H. Jullien, R. Hugues, and A. Rainoldi. 2011. "Prevention of Falling Risk in Elderly People: The Relevance of Muscular Strength and Symmetry of Lower Limbs in Postural Stability." *Journal of Strength and Conditioning Research* 25:567-574.

Reynolds, J.M., T.J. Gordon, and R.A. Robergs. 2006. "Prediction of One Repetition Maximum Strength From Multiple Repetition Maximum Testing and Anthropometry." *Journal of Strength Conditioning Research* 20:584-592.

Rosenberg, I.H. 1997. "Sarcopenia: Origins and Clinical Relevance." *The Journal of Nutrition* 127:990-991.

Shimokata, H., H. Shimada, S. Satake, N. Endo, K. Shibasaki, S. Ogawa, and H. Arai. 2018. "Chapter 2: Epidemiology of Sarcopenia." *Geriatrics and Gerontology International* 18:13-22.

Westcott, W.L. 2012. "Resistance Training is Medicine: Effects of Strength Training on Health." *Current Sports Medicine Reports* 11:209-216.

Whitehurst, M.A., B.L. Johnson, C.M. Parker, L.E. Brown, and A.M. Ford. 2005. "The Benefits of a Functional Exercise Circuit for Older Adults." *Journal of Strength and Conditioning Research* 19:647-651.

Zourdos, M.C., A. Klemp, C. Dolan, J.M. Quiles, K.A. Schau, E. Jo, E. Helms, et al. 2016. "Novel Resistance Training-Specific RPE Scale Measuring Repetitions in Reserve." *Journal of Strength and Conditioning Research* 30:267-275.

Lab 12

Campbell, S.C. 1982. "A Comparison of the Maximal Voluntary Ventilation With the Forced Expiratory Volume in One Second: An Assessment of Subject Cooperation." *Journal of Occupational Medicine* 24:531-533.

Coates, A.L., S.L. Wong, C. Tremblay, and J.L. Hankinson. 2016. "Reference Equations for Spirometry in the Canadian Population." *Annals of the American Thoracic Society* 13:833-841.

Gaensler, E.A. and G.W. Wright. 1966. "Evaluation of Respiratory Impairment." *Archives of Environmental Health* 12:146-189.

Gold, W.M. and L.L. Koth. 2016. "Pulmonary Function Testing." In *Murray and Nadel's Textbook of Respiratory Medicine.* 6th ed. Philadelphia, PA: Elsevier Saunders.

Haff, G.G. and C. Dumke. 2019. *Laboratory Manual for Exercise Physiology.* 2nd ed. Champaign, IL: Human Kinetics.

Johnson, J.D. and W.M. Theurer. 2014. "A Stepwise Approach to the Interpretation of Pulmonary Function Tests." *American Family Physician* 89:359-366.

MacNee, W. 2006. "ABC of Chronic Obstructive Pulmonary Disease: Pathology, Pathogenesis, and Pathophysiology." *British Medical Journal* 332:1202-1204.

Martinez-Pitre, P.J., B.R. Sabbula, and M. Cascella. 2021. "Restrictive Lung Disease." *StatPearls.* Treasure Island, FL: StatPearls Publishing.

Pellegrino, R., G. Viegi, V. Brusasco, R.O. Crapo, F. Burgos, R. Casaburi, A. Coates, et al. 2005. "Interpretative Strategies for Lung Function Tests." *European Respiratory Journal* 26:948-968.

Ranu, H., M. Wilde, and B. Madden. 2011. "Pulmonary Function Tests." *Ulster Medical Journal* 80:84-90.

Stocks, J. and P.H. Quanjer. 1995. "Reference Values for Residual Lung Volume, Functional Residual Capacity and Total Lung Capacity. ATS Workshop on Lung Volume Measurements. Official Statement of The European Respiratory Society." *The European Respiratory Journal* 8:492-506.

Lab 13

AlGhatrif, M. and J. Lindsay. 2012. "A Brief Review: History to Understand Fundamentals of Electrocardiography." *Journal of Community Hospital Internal Medicine Perspectives* 2:1-5. https://doi.org/10.3402/jchimp.v2i1.14383.

American College of Sports Medicine (ACSM). 2021. *ACSM's Guidelines for Exercise Testing and Prescription.* 11th ed. Baltimore: Lippincott Williams & Wilkins.

Dunbar, C. and B. Saul. 2009. *ECG Interpretation for the Clinical Exercise Physiologist.* 1st ed. Baltimore: Lippincott Williams & Wilkins.

Francis, J. 2016. "ECG Monitoring Leads and Special Leads." *Indian Pacing and Electrophysiology Journal* 16:92-95.

Goldberger, A., Z. Goldberger, and A. Shvilkin. 2013. *Goldberger's Clinical Electrocardiography: A Simplified Approach.* 8th ed. Philadelphia, PA: Elsevier Saunders.

Mason, R. and I. Likar. 1966. "A New System of Multiple-Lead Exercise Electrocardiography." *American Heart Journal* 71:196-205.

Scheidt, S. and J. Erlebacher. 1987. *Basic Electrocardiography.* 1st ed. Oxfordshire, UK: Taylor & Francis Group.

Thaler, M. 2018. *The Only EKG Book You'll Ever Need.* 9th ed. Baltimore: Lippincott Williams & Wilkins.

Lab 14

Ambrose, A.F., L. Cruz, and G. Paul. 2015. "Falls and Fractures: A Systematic Approach to Screening and Prevention." *Maturitas* 82:85-93.

American College of Sports Medicine (ACSM). 2021. *ACSM's Guidelines for Exercise Testing and Prescription.* 11th ed. Baltimore: Lippincott Williams & Wilkins.

Bergen, G., M.R. Stevens, and E.R. Burns. 2016. "Falls and Fall Injuries Among Adults Aged ≥65 Years: United States, 2014." *Morbidity and Mortality Weekly Report* 65:993-998.

Bohannon, R.W. and Y.C. Wang. 2019. "Four-Meter Gait Speed: Normative Values and Reliability Determined for Adults Participating in the NIH Toolbox Study." *Archives of Physical Medicine and Rehabilitation* 100:509-513.

Bortz, W.M. II. 2002. "A Conceptual Framework of Frailty: A Review." *The Journals of Gerontology: Series A* 57:M283-M288.

Bull, F.C., S.S. Al-Alsari, S. Biddle, K. Borodulin, M.P. Buman, G. Cardon, C. Carty, et al. 2020. "World Health Organization 2020 Guidelines on Physical Activity and Sedentary Behaviour." *British Journal of Sports Medicine* 54:1451-1462.

Campbell, A.J. and D.M. Buchner. 1997. "Unstable Disability and the Fluctuations of Frailty." *Age and Ageing* 26:315-318.

Cenzer, I.S., V. Tang, W.J. Boscardin, A.K. Smith, C. Ritchie, M.I. Wallhagen, R. Espaldon, and K.E. Covinsky. 2016. "One-Year Mortality After Hip Fracture: Development and Validation of a Prognostic Index." *Journal of the American Geriatrics Society* 64:1863-1868.

Edemekong, P.F., D.L. Bomgaars, S. Sukumaran, and S.B. Levy. 2021. "Activities of Daily Living." *StatPearls [Internet].*

Fritz, S. and M. Lusardi. 2009. "White Paper: Walking Speed: The Sixth Vital Sign." *Journal of Geriatric Physical Therapy* 32(2):2-5.

Gibson, A., D. Wagner, and V. Heyward. 2019. *Advanced Fitness Assessment and Exercise Prescription.* 8th ed. Champaign, IL: Human Kinetics.

Iverson, G.L. and M.S. Koehle. 2013. "Normative Data for the Balance Error Scoring System in Adults." *Rehabilitation Research and Practice* 2013:846418.

Kyrdalen, I.L., P. Thingstad, L. Sandvik, and H. Ormstad. 2019. "Associations Between Gait Speed and Well-Known Fall Risk Factors Among Community-Dwelling Older Adults." *Physiotherapy Research International* 24:e1743.

Middleton, A., S.L. Fritz, and M. Lusardi. 2015. "Walking Speed: The Functional Vital Sign." *Journal of Aging and Physical Activity* 23:314-322.

Nowak, A. and R.E. Hubbard. 2009. "Falls and Frailty: Lessons From Complex Systems." *Journal of the Royal Society of Medicine* 102:98-102.

Riemann, B.L., K.M. Guskiewicz, and E.W. Shields. 1999. "Relationship Between Clinical and Forceplate Measures of Postural Stability." *Journal of Sport Rehabilitation* 8:71-82.

Rikli, R.E. and C.J. Jones. 1997. "Assessing Physical Performance in Independent Older Adults: Issues And Guidelines." *Journal of Aging and Physical Activity* 5:244-261.

Rikli, R.E. and C.J. Jones. 1999. "Development and Validation of a Functional Fitness Test for Community-Residing Older Adults." *Journal of Aging and Physical Activity* 6:127-159.

Rikli, R. and C. Jones. 2013. *Senior Fitness Test Manual*. 2nd ed. Champaign, IL: Human Kinetics.

United States Census Bureau. 2018. "Older People Projected to Outnumber Children for the First Time in U.S. History." Last modified October 8, 2019. www.census.gov/newsroom/press-releases/2018/cb18-41-population-projections.html.

Appendix A

Nursing2023 Drug Handbook. 2023. 43rd ed. Baltimore: Lippincott Williams & Wilkins.

Appendix B

Liguori, G. and American College of Sports Medicine (ACSM). 2021. *ACSM's Guidelines for Exercise Testing and Prescription*. 11th ed. Philadelphia, PA: Wolters Kluwer.

Jeffrey M. Janot, PhD, ACSM CEP, is a professor and the chair of the department of kinesiology at the University of Wisconsin–Eau Claire. He primarily teaches in the rehabilitation science program, focusing on exercise physiology, fitness assessment and exercise prescription for healthy and clinical populations, and electrocardiography.

Janot received his PhD in health, physical education, and recreation, with a focus on exercise physiology, from the University of New Mexico. His research interests include sports physiology in hockey and baseball, applied cardiovascular physiology, and enhancing functional abilities of younger and older adults and those with chronic disease.

Janot holds certifications as an ACSM Clinical Exercise Physiologist and USA Hockey Level 5 Master Coach. Additionally, Janot has coached hockey at the youth level and currently coaches baseball at Eau Claire Memorial High School. He resides in Eau Claire, Wisconsin, with his wife, Jill, and their two children, Ben and Zach.

Nicholas M. Beltz, PhD, ACSM RCEP, is an assistant professor in kinesiology. He currently serves as the director of the Exercise Physiology Laboratory and director of the Community Fitness Program at the University of Wisconsin–Eau Claire. His teaching responsibilities are within the rehabilitation science and exercise science programs, specifically courses in laboratory methods, exercise physiology, and community fitness programming.

Beltz received his PhD in physical education, sports, and exercise science, with an emphasis in exercise science, from the University of New Mexico. His research is focused on central cardiovascular responses to concurrent exercise, evaluation of maximal oxygen uptake, and acute and chronic responses to resistance training.

Beltz has coached high school football and currently coaches baseball at Eau Claire Memorial High School. He resides in Eau Claire, Wisconsin, with his beautiful wife, Kjirsten, and their amazing daughters, Baylor and Bennett.